T0135543

From multiple senses to perceptual awareness

Dissertation

zur Erlangung des Grades eines
Doktors der Naturwissenschaften

der Mathematisch-Naturwissenschaftlichen Fakultät
und
der Medizinischen Fakultät
der Eberhard-Karls-Universität Tübingen

vorgelegt
von

Anette S. Giani
aus Aachen, Deutschland

April - 2013

Bibliografische Information der Deutschen Nationalbibliothek

Die Deutsche Nationalbibliothek verzeichnet diese Publikation in der
Deutschen Nationalbibliografie; detaillierte bibliografische Daten sind
im Internet über http://dnb.d-nb.de abrufbar.

ISBN 978-3-8325-3656-5

Logos Verlag Berlin GmbH
Comeniushof, Gubener Str. 47,
10243 Berlin
Tel.: +49 (0)30 42 85 10 90
Fax: +49 (0)30 42 85 10 92
INTERNET: http://www.logos-verlag.de

Tag der mündlichen Prüfung: 17.01.2014

Dekan der Math.-Nat. Fakultät:	Prof. Dr. W. Rosenstiel
Dekan der Medizinischen Fakultät:	Prof. Dr. I. B. Autenrieth
1. Berichterstatter:	Prof. Dr. U. Noppeney
2. Berichterstatter:	Prof. Dr. C. Braun
Prüfungskommission:	Prof. Dr. C. Braun
	Prof. Dr. A.J. Fallgatter
	Prof. Dr. H. Preissl
	PD Dr. A. Bartels

TABLE OF CONTENTS

INTRODUCTION

"Reality is just an illusion, albeit a very persistent one."

Albert Einstein

Throughout the day, our senses provide us with a rich stream of information about the environment: We see colours and shapes, hear music or smell food. With seemingly no effort, the human brain integrates these signals to create a conscious sensory experience of the external world. Yet, this sensory experience is not a truthful representation of the physical world. Instead, our perception is crucially shaped by a variety of processes, two of which are the focus of the current work: multisensory integration and awareness.

At any time, our sensory receptors receive visual, tactile, auditory or other sensory signals. These signals can be more or less reliable, such as vision during day and night. Fortunately, we can interpret even noisy signals mainly because we are equipped with multiple senses. If fog impedes sight, for example, we can rely on auditory or tactile cues to orient ourselves. Yet, to profit from our various senses, sensory information needs to be combined, a process called 'multisensory integration' (or multimodal integration). Multisensory integration does not only allow us to interpret ambiguous situations (van Ee et al., 2009; Battaglia et al., 2010), but also speeds up reaction (Molholm et al., 2002; Diederich and Colonius, 2004; Van der Burg et al., 2008), enhances performance (e.g.: Van der Burg et al., 2008) and improves learning (e.g.: Shams and Seitz, 2008).

Since multisensory influences imprint themselves so naturally on us, we usually do not become aware of how these influences alter our perception. For example, in a famous experiment, McGurk and McDonald (1976) showed that seeing lip movements can change our perception of speech sounds. Similarly, visual stimulus can influence the perceived location of a simultaneously occurring sound (e.g.: Bertelson and Radeau, 1981).

7

At any given time, a massive variety of visual, somatosensory and auditory signals reach our sensory receptors. (Think, for example, of standing on a crowed street). Counterintuitively, only a fraction of this sensory information eventually reaches our awareness. Most sensory signals are supressed to prevent overload. The impact of information selection on perception is nicely illustrated in change blindness. Observers may fail to detect even large environmental changes such as two people on a photograph exchanging heads (reviewed in: Simons and Levin, 1997; Simons and Rensink, 2005). Yet, even though we remain unaware of most details, these details may nevertheless impact our behaviour. Cortically blind patients cannot consciously see visual objects. Nevertheless, they might localize them within space; a phenomenon called "blindsight" (Poppel et al., 1973).

Multisensory integration and awareness are constant companions of our daily life which determine how we perceive the world. In contrast to their impact, relatively little is known about the mechanisms that enable perceptual awareness within a multisensory world. For example, does multisensory integration occur automatically or are higher order cognitive processes (such as awareness) necessary to bind the information? And where does awareness emerge within the human brain? After a short introduction on the fundamental aspects of multisensory integration and awareness, three experimental studies designed to provide further insights on auditory and visual perception are presented.

MULTISENSORY PERCEPTION

WHY DOES A VENTRILOQUIST'S DUMMY TALK? - CONSEQUENCES OF MULTISENSORY INTEGRATION

Since multisensory perception occurs so naturally in daily live, its complexity becomes evident only when our perception is deceived. For example, misleading (i.e. spatially conflicting) audiovisual signals are responsible for the ventriloquist illusion, i.e. the perception of a talking dummy. The co-occurring ventriloquist's speech sounds and the dummy's mouth movements are integrated into one coherent percept. In doing so, the perceived location of one sensory input (e.g. auditory) is shifted towards the location of a temporally correlated but spatially displaced input of another sensory modality (e.g. visual) (e.g.: Bermant and Welch, 1976; Bertelson and Radeau, 1981). This illusion is extremely persistent. We even experience it while watching TV: The actors' speech sounds are perceived as being co-localized with the actors' faces even though they actually arise from the speakers. Accordingly, it is believed that the ventriloquist illusion occurs automatically, i.e. independent of higher cognitive processes (Bertelson and Aschersleben, 1998; Bertelson et al., 2000b; Vroomen et al., 2001).

During the spatial ventriloquist illusion, visual signals commonly capture the auditory signals' perceived location. Similarly, vision can also shape vestibular information, e.g. during vection. Vection refers to the illusionary sensation of self-motion induced by optical flow patterns (e.g.: Hettinger et al., 1990). For instance, sitting in a train we may incorrectly perceive our wagon to move off because we see another train passing by the window. However, it is not necessarily the case that the visual signal dominates the other senses' signals. Instead, the way multisensory signals influence each another depends on each stimulus' reliability (Alais and Burr, 2004; Ernst and Bulthoff, 2004). While vision usually provides more reliable estimates of spatial cue location, audition generally provides more reliable estimates of temporal location. Consequently, audition influences our perception of a visual stimulus' onset. During the "flash illusion", for example, visual information is shaped by audition (Shams et al., 2000). One single light flash is incorrectly perceived as two light flashes if it is paired with two sound bursts. Similarly, during 'auditory driving' the perceived flickering rate of a visual stimulus is adapted by the flutter frequency of a co-occurring sound (Regan and Spekreijse, 1977; Noesselt et al., 2008). Remarkably, auditory signals can even shape tactile perception. Tactile sensations during hand-rubbing are shaped by sounds, inducing the parchment skin effect (Jousmaki and Hari, 1998).

Commonly, synchronicity forms the basis of the illusions described above. Co-occurring signals usually originate from one single object. Accordingly, multisensory stimuli that are co-localized in space and time are likely to be perceived as belonging to the same object (i.e. assumption of unity) even if both stimuli are completely unrelated otherwise (Welch and Warren, 1980). For instance, Van der Burg et al. (2008) show that a non-spatialized auditory

9

cue can help to guide visual search towards a simultaneously changing visual target. Interestingly, audiovisual cues are perceived as simultaneous even if they are physically separated by more than 100 ms (Dixon and Spitz, 1980). The brain can thus incorporate difference in audiovisual onset asynchronies that arise due to differences in travelling time in air and transduction rates within the brain (Spence and Squire, 2003).

Multisensory illusions elucidate the complexity and consequences of the integration process. The puzzle is that the brain integrates information with seemingly no effort. With the advent of sophisticated neuroimaging techniques, such as functional magnetic resonance imaging (fMRI) and magnetoencephalography (MEG), it has become possible to observe the neural processes underlying multisensory processes.

METHODOLOGICAL CONSIDERATIONS

The consequences of multisensory integration are readily visible at the behavioural level: Observers misperceive the location of a sound, they react faster or more accurately. However, visualizing multisensory benefits within the brain can be challenging.

Studying multisensory integration at the level of individual neurons is still relatively straightforward. Depending on whether a neuron responds to the stimuli of only one or multiple sensory modalities, it is classified as *unimodal* or *multimodal*, respectively. Importantly however, multimodal neurons need not necessarily integrate information across modalities. In fact, only a subset of multimodal neurons shows multisensory effects (Schroeder et al., 2001; Schroeder and Foxe, 2002; Stein and Stanford, 2008). To assess multisensory effects of single neurons, researchers traditionally compare the response of a multisensory stimulus to the responses of the component, unisensory stimuli. According to the '*max criterion*' a neuron integrates multisensory signals if a multisensory stimulus evokes more (or less) activity than its most (or least) effective component stimulus. This neural response pattern is called *multisensory enhancement* (or *depression*). Sometimes, this response enhancement even exceeds the summed activity evoked by the unisensory stimuli (i.e. '*additive criterion*'). In that case, bimodal neurons are further classified as *superadditive* (or *subadditive*) (**Figure 1**) (for comprehensive reviews see: Stein and Stanford, 2008; James and Stevenson, 2012).

To noninvasively visualize integration within the *human* brain researchers mostly rely on neuroimaging techniques, such as functional magnetic resonance imaging (fMRI), magnetoencephalography (MEG), or electroencephalography (EEG). Commonly, these imaging techniques describe the summed activity produced by millions of neurons, leaving the characteristics of the underlying populations of neurons unknown (Hillebrand and Barnes, 2002; Logothetis and Wandell, 2004). However, the measured response to an audiovisual stimulus differs depending on whether the underlying neural population is (1) unimodal auditory, (2) unimodal visual or (3) bimodal audiovisual. The task of interpreting neuroimaging data is further complicated by the fact that they mostly reflect the summed activity of a mixture of unimodal neurons from different modalities, in addition to bimodal neurons of which only some show multisensory enhancements (James and Stevenson, 2012). In such cases, multisensory stimuli necessarily evoke stronger (fMRI, M/EEG, PET) activity than the strongest of its component stimuli, because multisensory stimuli activate larger pools of neurons. Hence, the max criterion might be inadequate to assess multisensory integration in functional imaging data (**Figure 1**).

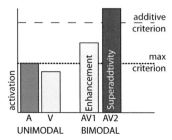

Figure 1: Classification of hypothetical neural responses to auditory (A, green), visual (V, yellow) and bimodal stimulation (AV, grey). The bimodal activations can be classified as a sensory enhancement (AV1, light grey) or superadditive (AV2, dark grey). Dashed and dotted lines represent the additive and max criterion, respectively. Sensory enhancement is classified as a multisensory effect according to the max, but not according to the additive criterion.

Alternatively, additivity may serve as a criterion for multisensory integration, e.g., by comparing the summed response for auditory (A) and visual (V) stimuli to the response for the compound audiovisual stimulus (AV ≠ A + V) (Calvert, 2001). In functional neuroimaging, however, this criterion is known to be extremely conservative and may lead to false negatives (Beauchamp, 2005; Goebel and van Atteveldt, 2009). For instance, showing superadditivity at a population level is particularly challenging since only some of the underlying neurons show multisensory effects. Moreover, according to the law of **"inverse effectiveness"**, the multisensory gain increases as the stimuli intensity decreases. Therefore, superadditivity might only be observable for weak stimuli (Stein and Stanford, 2008). Most problematic, however, are non-specific stimulus evoked processes such as arousal, anticipatory processes, etc. While these processes are counted twice for the sum of the two unisensory stimuli, they are counted only once for the compounded audiovisual stimulus (Teder-Salejarvi et al., 2002; James and Stevenson, 2012).

Several solutions have been proposed to address these problems (Teder-Salejarvi et al., 2002; Goebel and van Atteveldt, 2009; James and Stevenson, 2012). First, introducing a baseline condition can help to avoid confounds of non-specific stimulus evoked processes, for example when comparing audiovisual (AV) to auditory (A) and visual (V) responses. Testing specifically whether (AV + baseline) ≠ (A + V), non-specific confounds are counted twice for the summed response as well as for the compound stimuli. Beyond, multisensory integration may be identified as the relative difference between conditions. Making use of the law of inverse effectiveness, for example, one may compare the multisensory effect evoked by weak relative to strong stimuli. The corresponding comparison $(AV - (A + V))_{weak} \neq (AV - (A + V))_{strong}$ no longer depends on the baseline (James and Stevenson, 2012).

Nevertheless, these solutions may be too stringent or simply not practicable. Therefore, many studies fail to apply these solutions (Giard and Peronnet, 1999; Fort et al., 2002; Molholm et al., 2002; Beauchamp, 2005; Talsma and Woldorff, 2005; Raij et al., 2010). They therefore run the risk of reporting false positives, and reporting false positives can have serious consequences (Bennett et al., 2009). Finding new, easy and practicable criteria to assess multisensory integration is therefore crucial.

MULTISENSORY INTEGRATION WITHIN THE HUMAN BRAIN

With these considerations in mind, we can focus on the advances in multisensory brain research achieved in recent decades. Where and how does the human brain merge the various information sources into one coherent percept?

Perception was traditionally studied within a single modality. This way of studying perception advanced the view that multisensory interplay happens in so-called convergence areas, i.e. in regions receiving input from multiple areas (Meredith and Stein, 1986; Driver and Noesselt, 2008). The superior colliculus (SC) constitutes a classical convergence zone which receives visual, auditory, vestibular, proprioceptive and somatosensory inputs. Not surprisingly, the SC contains a large number of multimodal neurons (about 49 % in deep laminae) (Meredith and Stein, 1986). Moreover, pioneering work by Meredith and Stein (1986) established that, relative to unisensory stimuli, multisensory stimulus enhance responses of 45% of the multimodal neurons, while it depresses the response of about 20%. Strikingly, the mean response enhancement in firing rate, evoked by the multisensory stimuli, amounted to 1207%. Hence, the SC is firmly established as a location for multisensory stimuli integration.

Beyond subcortical areas, multisensory convergence zones also include higher order cortical areas. These include the superior temporal sulcus (STS) and parietal and frontal cortices (Bruce et al., 1981; Calvert et al., 2001; Beauchamp et al., 2004; van Atteveldt et al., 2004; Schlack et al., 2005; Noesselt et al., 2007; Werner and Noppeney, 2010). For example, early studies on animal neurophysiology showed that more than half of all neurons in STS receive inputs from at least two modalities (Bruce et al., 1981; reviewed in: Ghazanfar and Schroeder, 2006). Similarly, Schlack et al. (2005) observed that receptive fields of auditory and visual neurons overlap within the ventral intraparietal area (VIP). The authors therefore assume that VIP is involved in maintaining a coherent percept, independent from modality specific reference frames (Schlack et al., 2005).

However, more recently decades, a vast amount of research has challenged traditional views on multisensory integration. Nowadays it seems that multisensory integration happens essentially all along the sensory pathways (Ghazanfar and Schroeder, 2006). Even cortical regions that were once believed to be sensory specific can integrate multisensory signals (Calvert et al., 1997; Kayser et al., 2007). Kayser and colleagues (2007), for example, showed that visual stimuli can activate even the primary auditory cortex. Moreover, activity in auditory core and belt areas increased for audiovisual relative to auditory or visual stimuli. Moreover, evidence from MEG and EEG further suggested that multisensory integration can occur extremely early, starting around 50 ms after stimulus onset (Schroger and Widmann, 1998; Giard and Peronnet, 1999; Fort et al., 2002; Molholm et al., 2002; Murray et al., 2005; Talsma et al., 2007). (Interestingly though, those studies that controlled for nonspecific confounds instead found later multisensory effects (Teder-Salejarvi et al., 2002; Talsma and Woldorff, 2005).)

It remains controversial, though, whether or not these early activations can be explained by direct, feedforward connections. Laminar profiles of somatosensory responses in the auditory association cortex show that somatosensory input is conveyed by feedforward projections (Schroeder and Foxe, 2002). In line, evidence for early multisensory effects further supports

14

the idea of feedforward integration (Murray et al., 2005; Talsma et al., 2007). Lastly, anatomical connections between early, sensory cortices have been found supporting the idea of direct exchange of information between sensory specific cortices (Miller and Vogt, 1984; Falchier et al., 2002).

In spite of this evidence for direct feedforward connections, feedback connections from convergence areas are probably at least as important (Schroeder and Foxe, 2002). It appears that multisensory integration starts early within the hierarchy and increases during processing. Kayser and colleagues (2012) even claim that the activity evoked by multisensory stimuli roughly doubles from one processing stage to the next within the cortical hierarchy.

PERCEPTUAL AWARENESS

CONSEQUENCES OF AWARENESS

Our multiple senses are continuously receiving sensory signals. Yet, counterintuitively, the brain uses only a fraction of these signals to form a coherent picture of the world. Irrelevant information is supressed from awareness thus preventing perceptual overload. These limits in processing capacities are reflected in perceptual illusions such as change blindness (Simons and Levin, 1997; Simons and Rensink, 2005).

However, what happens to those signals that hit our sensory receptors but never reach the level of awareness? For instance, can unaware sensory information still influence our behaviour? The latter question achieved public attention in 1957. At that time, James Vicary pretended that unconsciously viewed advertisements increase sales. Even though it turned out that Vicary had falsified his results, some have believed in the power of unconscious perception ever since. Consequently, dubious videos can be purchased that include subliminal messages. Allegedly, watching these videos should help the observer to lose weight, reduce stress or gain confidence (see: www.mindmaster.tv).

Even though it is unlikely that subliminal messages increase confidence, unaware sensory signals nevertheless affect behaviour. Strahan and colleagues (2002), for example, showed that subliminal messages can prime cognition if the prime is goal-relevant. In particular, they showed that thirst-related primes caused participants to consume more liquid only if participants were thirsty.

The effect of unaware information on behaviour can be observed even more strikingly during blindsight. Blindsight allows visual information to guide reaching movements in the absence of any conscious visual experience (Weiskrantz et al., 1974). Importantly, blindsight-like phenomena depend on an intact retina. From there, visual information enters the visual pathway and probably reaches the cortex via thalamo-cortical projections (surpassing the lesioned visual cortex). Commonly, these phenomena are interpreted as showing that even unaware information is processed within the brain. Knowing the neural correlates of awareness would boost our understanding of how awareness influences cognition. Accordingly, for decades, neuroscientists have been interested in how awareness emerges within the brain.

STUDYING AWARENESS: METHODOLOGICAL IMPLICATIONS

Studying perceptual awareness is not trivial for a number of reasons. First, awareness has not been defined unambiguously. It may relate to states of vigilance (such as being awake or under hypnosis) or to the specific content of an experience, e.g. hearing a certain tone or seeing a colour (Chalmers, 2000; Dehaene and Changeux, 2011). Philosophical discussions surround experimental research on "the neural correlates of consciousness" (Chalmers, 2000). To sidestep these debates, we define perceptual awareness operationally, as the availability of a stimulus' content. Hence, being aware of a grating allows participants to report its presence either verbally or behaviourally (Dehaene and Changeux, 2011). We therefore assume that phenomenal awareness correlates with the availability of its content. Theoretically, it may be possible for participants to have phenomenal awareness, even though they remain unable to report its content. For more philosophical discussions on that topic see, for example, Chalmers (1995).

Importantly, there are at least two criteria to probe availability of a phenomenal awareness: we can rely on a participant's subjective report (subjective criterion) or we can test whether a

16

participant can correctly identify the stimulus content (objective criterion). For instance, if we would like to know if someone is aware of a grating, we can either ask him/her whether or not he/she perceived the tone or we can ask whether the grating was tilted to the left or to the right. Notably, blindsight-like vision would be classified as unconscious vision only according to subjective criteria. Alternatively, using objective criteria, it would be classified as conscious vision, even though the patient claims that he/she had no visual awareness.

Beyond definitional issues, it remains challenging to develop suitable paradigms and stimuli to study awareness. Traditionally, two experimental conditions are contrasted to detect changes in behaviour or neural activity. Crucially, only the variable of interest (i.e. awareness) is allowed to vary between conditions. This means that physically identical stimuli have to evoke different states of awareness. Otherwise, low level influences of stimuli's identity cannot be disentangled from awareness effects.

During the last decades, many paradigms have been developed to study awareness while keeping the physical stimulus parameters constant. During binocular rivalry, for example, two dissimilar images are presented to each of the two eyes. These images compete for access to consciousness. As a result, alternatingly either one or the other image is dominant and will become aware. Hence, perceptual awareness changes every few seconds allowing researchers to study its underlying neural bases (Tong et al., 2006).

A variant of binocular rivalry, called continuous flash suppression (CFS), allows gaining more control over dominance rates. As for binocular rivalry, two dissimilar images are displayed to each of the two eyes. However, in CFS one image contains a rapidly changing, high-contrast pattern. This pattern continuously supresses the other, static, low-contrast image from awareness (Tsuchiya and Koch, 2005). CFS thus renders a visual stimulus unaware for an extended period of time. There are many more paradigms that allow studying visual awareness. These include forward and backward masking, attentional blink, subliminal perception or change blindness. In contrast, only recently, scientists started to develop equivalent paradigms to study auditory awareness. Of those, informational masking has proven powerful tool to study auditory awareness within complex auditory scenes. In informational masking experiments, participants typically have to detect a sequence of pure-tone targets embedded within a multi-tone mask (Neff and Green, 1987). Stimulus detection thus requires participants to identify the sequence of target tones via temporal integration and stream segregation. Critically, target tones are presented in a protected region in frequency space thus avoiding overlap of sound energies of masking and target tones (Durlach et al., 2003). Consequently, target tones remain well above detection threshold evoking two clearly distinct percepts: Either they become clearly audible or they remain completely unaware (Wiegand and Gutschalk, 2012). Identical physical stimuli thus lead to different conscious reports that should be reflected within cortical processing.

In conclusion, powerful experimental designs allow neuroscientists to study the neural processes underlying perceptual awareness. However, since awareness remains an ambiguous term, clear working definitions should be formulated. Moreover, design and stimuli need to be developed extremely carefully.

AWARENESS WITHIN THE HUMAN BRAIN

Around 1600, René Descartes formulated one of the first theories on the neural correlates of awareness. He proposed that mind and body interact via the pineal gland (Van Gijn, 2005). Today, most neuroscientists believe that awareness is an emerging property of the human brain. Therefore, with the advent of modern neuroimaging techniques, they aimed at visualizing the neural processes underlying awareness. Surprisingly however, no coherent picture emerged (Dehaene et al., 2006).

It remains debated, for instance, whether or not early sensory processing can give rise to perceptual experiences (Dehaene et al., 2006). The necessity of early visual cortex for conscious experience was demonstrated by Cowey and colleagues (2000). They showed that transcranial magnetic stimulation (TMS) can induce phosphenes in blind people only if early visual cortex is intact (Cowey and Walsh, 2000). Accordingly, lesions to primary visual cortex lead to the loss of all phenomenal visual experience. Further, neuroimaging studies identified an association between visual awareness and early cortical processing (Pins and Ffytche, 2003; Ress and Heeger, 2003; Gutschalk et al., 2008).

18

Collectively, these results suggest that early visual processing is necessary for perceptual awareness. However, is it sufficient to evoke awareness, too? Pascual-Leone and Walsh (2001) showed that TMS-induced moving phosphenes can be eliminated by temporarily deactivating the visual cortex. This result indicates that back projections to V1, rather than its mere activation, are correlated with visual awareness. Moreover, most M/EEG studies found that later (> 150 ms) components correlate with perceptual awareness (Sergent et al., 2005; Del Cul et al., 2007; Bekinschtein et al., 2009; Brancucci et al., 2011; Shen and Alain, 2011; Auksztulewicz et al., 2012). For example, Sergent and colleagues (2005) showed that early EEG components are not affected by awareness. In contrast, awareness was related to a late parieto-frontal network. Similarly, Del Cul et al. (2007) identified highly distributed fronto-parieto-temporal activations as a correlate of awareness. This activation started at around 300 ms. Evidence from fMRI studies further supports the crucial role of higher level processing for awareness (Haynes et al., 2005). Lastly, Panagiotaropoulos and colleagues (2012) observed that the neural activity in lateral prefrontal cortex correlates with phenomenal awareness.

In their review Dehaene and colleagues (2006) merge the apparently contradictory evidence and propose a taxonomy on the basis of a 'global neuronal workspace hypothesis'. In particular, they reason that early cortical activations are an indication of subliminal or preconscious processing, while access to consciousness depends on recurrent activation within a global workspace network, including parietal and frontal cortices. Within the global workspace network sensory information needs to become functionally integrated. In their review Engel and Singer (2001) stress the importance of synchronization of neuronal discharges as the primary agent for perceptual binding. In particular, they reason that synchronization is "gating the access to consciousness". Accordingly, Fries et al. (1997) showed that perceptual awareness relates to synchronized discharges within the gamma band frequency. Collectively, it appears that (gamma band) oscillations and phase synchronizations within a global workspace are crucial to form a coherent, conscious percept (Rodriguez et al., 1999; Engel and Singer, 2001; Melloni et al., 2007; Doesburg et al., 2009).

PUTTING THE PUZZLE TOGETHER: MULTISENSORY AWARENESS

The present work comprises three experiments, investigating the (neural) mechanisms underlying conscious auditory and visual perception. In **chapter 1** we characterize the relation between multisensory integration and perceptual awareness. Conventional views assume that integration happens automatically, i.e. in the absence of attention or awareness (Bertelson et al., 2000a; Vroomen et al., 2001). Combining spatial ventriloquism and continuous flash suppression, we specifically tested for an influence of awareness on audiovisual spatial integration.

Following, we were interested in the neural dynamics underlying conscious, auditory and visual perception. To assess neural activity we used Magnetoencephalography (MEG). However, as discussed above, identifying multisensory integration remains a major challenge to neuroimaging. In **chapter 2** we therefore describe a frequency based technique, developed to unambiguously index sensory integration (Regan and Regan, 1988; Regan et al., 1995). If successful, this technique offers a unique opportunity to image the effects of awareness on multisensory integration.

Lastly, in **chapter 3** we looked at the neural dynamics that enable auditory awareness of a pair of target tones. Combining MEG, informational masking and dynamic causal modelling (DCM) we modelled complex cortical interactions that gave rise to awareness. Future work will build upon those results to understand even more complicated audiovisual interactions and their relation to awareness.

CHAPTER 1:

THE INVISIBLE VENTRILOQUIST

Anette S. Giani[1], Verena Conrad[1], Masataka Watanabe[1] & Uta Noppeney[1,3]

[1] Max Planck Institute for Biological Cybernetics, Tübingen, Germany

[3] Computational Neuroscience and Cognitive Robotics Centre, University of Birmingham, UK

ABSTRACT

Multisensory integration is fundamental for effective interactions with our natural environment. However, it remains controversial whether integration of sensory signals is automatic or dependent on higher cognitive processes. Past research has demonstrated that ventriloquism (a perceptual illusion that illustrates how the brain integrates sensory signals for spatial orienting) is unaffected by attentional and decisional biases. Yet, the role of awareness in this process is unknown. To investigate whether perceptual awareness shapes integration of sensory signals into spatial representations, we combined spatial ventriloquism with dynamic continuous flash suppression. Critically, as flash suppression obliterated visual awareness only in a fraction of trials, we were able to quantify the effect of awareness on multisensory integration by comparing spatial ventriloquism for physically identical flashes that were visible or invisible. Challenging conventional views of automatic spatial integration, our results demonstrate that spatial ventriloquism is profoundly modulated by perceptual awareness. While a strong ventriloquist effect was observed for visible flashes, it was nearly abolished when the flash is invisible. Nevertheless, we observed a small but robust ventriloquist effect, even when participants were at chance when locating the visual flash. This indicates that visual stimuli that evade perceptual awareness can alter the perceived location of auditory signals. Collectively, our results call for reconsideration of the relation between multisensory integration, attention and awareness. They suggest that sensory signals are integrated into coherent spatial representations at pre- and post-aware processing stages that are potentially implemented in distinct neural circuitries.

INTRODUCTION

Information integration is critical for effective interactions with our natural environment. To form a coherent and more reliable percept, the brain needs to integrate signals from multiple senses. However, it remains controversial, to what extent multisensory integration is automatic or dependent on higher cognitive processes such as attention or awareness (Talsma et al., 2010).

While integration of higher order features (e.g. phonology) has been shown to rely on attention and awareness (Alsius et al., 2005; Munhall et al., 2009), low level spatiotemporal integration is thought to be immune to decisional and attentional biases (Bertelson and Aschersleben, 1998; Bertelson et al., 2000a; Vroomen et al., 2001). Evidence for this comes, most prominently, from spatial ventriloquism, a perceptual illusion that emerges when sensory signals are artificially brought into spatial conflict thereby exemplifying how the brain integrates sensory signals into a coherent spatial representations (Radeau and Bertelson, 1977; Bertelson and Radeau, 1981). In spatial ventriloquism, the perceived location of one sensory input (e.g. auditory) is shifted towards the location of a temporally correlated but spatially displaced input of another sensory modality (e.g. visual) and vice versa depending on the relative sensory reliabilities (Alais and Burr, 2004).

Critically, ventriloquism has been observed, even when decisional biases and response strategies are carefully controlled (Vroomen and de Gelder, 2004). Moreover, spatial ventriloquism has been shown to be unaffected by endogenous or exogenous spatial attention (Bertelson et al., 2000a; Vroomen et al., 2001). In fact, it is thought to facilitate and occur at least to some extent prior to spatial attentional selection (Driver, 1996). Likewise, ventriloquism was not influenced by modality-specific attention, i.e. whether participants focused on a particular sensory modality (Vroomen and de Gelder, 2004).

Collectively, this body of research suggests that spatial ventriloquism emerges at the sensory processing level largely unaffected by attentional or decisional control. Therefore, one may ask whether it emerges automatically, i.e. even prior to participants' awareness. Given the absence of attentional effects and past research showing that attention can modulate signal

23

processing even prior to perceptual awareness (Watanabe et al., 2011), one would expect only negligible influences of awareness on ventriloquism. Indeed, initial tentative evidence from patients with spatial hemineglect suggests that audiovisual spatial ventriloquism persists for visual signals that participants are not aware of (Bertelson et al., 2000b). Yet, these results need to be interpreted with caution, as the ventriloquist effect was reported as significant only for visual signals in patients' neglected, but not in their intact hemifield. Furthermore, this study characterized the ventriloquist effect only for unaware but not for aware visual signals in patients' neglected hemifield. Therefore, it could not formally quantify the contributions of awareness to audiovisual spatial integration.

To investigate whether integration of auditory and visual signals into multisensory spatial representations depends on perceptual awareness, the present study combined spatial ventriloquism with dynamic continuous flash suppression (CFS) (Tsuchiya and Koch, 2005; Maruya et al., 2008). Dynamic CFS is a novel technique that suppresses participants' awareness of monocularly viewed events by simultaneously presenting rapidly changing masks to the other eye (Maruya et al., 2008). Using CFS we presented participants' suppressed eye with a brief visual flash to their left or right hemifield. In synchrony with the flash, a brief beep was played in the centre, left or right hemifield. Critically, we selected the saliency of the visual flash, such that the dynamic flash suppression obliterated visual awareness only in a fraction of trials. Comparing spatial ventriloquism for visible and invisible - yet physically identical - flashes allowed us to quantify the effect of awareness on multisensory integration. Challenging conventional views of automatic spatial integration, our results demonstrate that spatial ventriloquism is profoundly modulated by perceptual awareness. Nevertheless, we observed a small but significant ventriloquist effect for flashes that are invisible according to subjective and objective awareness criteria. These findings call for reconsideration of the relation between multisensory integration, attention and awareness.

24

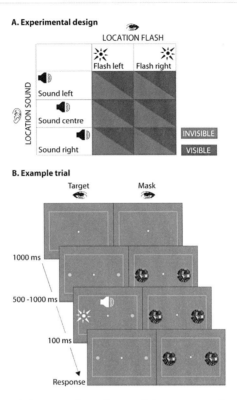

Figure 1. Experimental design and conditions. (*A*) Experimental paradigm: 2 x 3 x 2 factorial design with the factors: (1) Flash location: left, right; (2) Sound location: left, centre, right, (3) Visibility: Visible, Invisible. (**B**) Schematic time flow: Participants viewed one half of the monitor with each eye. One eye viewed the target stimuli, while the other viewed the CFS masking stimuli. On each trial, either the left or the right target's colour changed to white ('flash'). Simultaneously, a sound occurred at the left, right or centre.

25

RESULTS

In a ventriloquist paradigm, participants were presented with an auditory beep emanating from one of three potential locations: left, centre, right (**Figure 1**). In synchrony with the beep, one eye was presented with a brief flash either in participants' left or right hemifield. Participants' awareness to these flashes was suppressed by showing dynamic Mondrians to the other eye (i.e. dynamic flash suppression) (Maruya et al., 2008). For each trial, participants reported the location of the beep (left, centre, right) and rated the visibility of the flash (visible, unsure, invisible). This visibility judgment provides a 'subjective awareness criterion'. Critically, the flash was visible only in a fraction of trials allowing us to quantify the effect of awareness on multisensory integration by comparing spatial ventriloquism for visible and invisible - yet physically identical - flashes. In addition, on 22.2 % of the trials, participants were also asked to locate the flash. This allowed us to assess the spatial information that is available for visual spatial localization during visible and invisible trials. Furthermore, we were able to select participants that were better than chance when locating visible flashes (i.e. 28 'reliable participants' included in total in the study) and not better than chance when locating invisible flashes. The latter allowed us to investigate the influence of flashes that participants cannot locate and were thus not aware of in an objective sense on sound localization (i.e. objective awareness criterion, see analysis 3).

Analysis 1 – Effect of subjective visibility on spatial ventriloquism:

First, we investigated whether participants' awareness (i.e. the visibility of the flash) modulated spatial ventriloquism, i.e. how participants integrated auditory and visual signals into spatial representations. To ensure reliable parameter estimation, this analysis included only the 11 participants that had at least 10 trials per condition and visibility level. Indeed, at the group level, the visibility judgments limited to this group were relatively balanced with 35.5 % of the trials rated as visible, 15.7 % as unsure and 48.8 % as invisible. The 'unsure' response option was primarily included to encourage participants to categorize trials as invisible when they were fully unaware of the flash. It was not of further interest for the analysis. Hence, we entered the perceived auditory location for visible and invisible trials only into a 2 x 3 x 2 repeated measures ANOVA with the factors (1) flash location (left, right), (2) sound location (left, centre, right) and (3) flash visibility (visible, invisible). Not surprisingly, the repeated measures ANOVA showed a significant main effect of true sound location on perceived sound location ($F_{(1.04, 10.43)}$ = 49.02, p < 0.001). More importantly, it also identified a significant main effect of flash location ($F_{(1, 10)}$ = 13.45, p = 0.004) and a flash location x flash visibility interaction ($F_{(1, 10)}$ = 12.55, p = 0.005). The significant interaction demonstrates that the visual flash influences the perceived sound location depending on whether or not participants are aware of the flash (**Figure 2A**). To further investigate the effect of the flash on perceived sound location, we preformed two post-hoc 2 (visual flash: left, right) x 3 (sound location: left, centre, right) repeated measures ANOVAs, separately for visible and invisible trials. Replicating the well-established ventriloquist effect for visible trials, the analysis identified a highly significant main effect of both true sound location ($F_{(1.08, 10.76)}$ = 33.23, p < 0.001) and flash location ($F_{(1, 10)}$ = 13.10, p = 0.005) on perceived sound location, when participants were aware of the visual flash. For invisible trials, we observed a main effect of true sound location ($F_{(1.02, 10.24)}$ = 56.84, p < 0.001), but only a marginally significant main effect of flash location ($F_{(1, 10)}$ = 3.71, p = 0.083). However, as the interaction between sound and flash location was not significant ($F_{(1.22, 12.19)}$ = 0.22, p = 0.697), we were able to average the perceived sound location across the three levels of true sound location separately for left and right flashes. In line with previous research on spatial ventriloquism, we hypothesized that on average the perceived sound location would be shifted towards the right for right relative to left flashes. Hence, we entered the perceived sound location for right and left flashes into a directional paired t-test. This paired-t-test identified that the perceived sound location was significantly nudged to the right for right relative to left visual flashes ($t_{(10)}$ = 1.93, p = 0.042) (**Figure 2B**).

In summary, analysis 1 identified a highly significant interaction between flash location and flash visibility. In other words, physically identical flashes induced a shift in perceived sound location (i.e. ventriloquist effect) that depended on whether or not they were consciously perceived. Separate post hoc ANOVAs confirmed a highly reliable ventriloquist effect for visible flashes and a small but significant ventriloquist effect for invisible flashes.

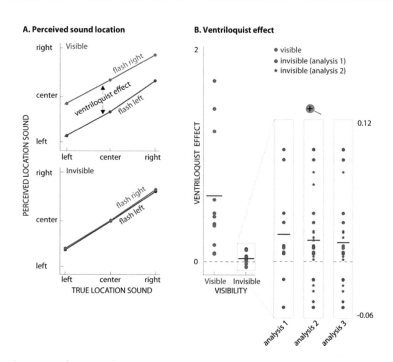

Figure 2. The ventriloquist effect. (*A*) Effect of awareness on spatial ventriloquism. Perceived sound location as a function of true sound location (left, centre, right) and flash (left, right) shown separately for visible (top, blue) and invisible (bottom, green) trials from analysis 1. (*B*) The ventriloquist effect averaged across all sound locations shown for visible (blue) and invisible (green) trials. The enlarged columns show the significant ventriloquist effect separately for the three analyses that defined awareness via subjective (analysis 1 and 2) and objective criteria (analysis 3).

Analysis 2 – Spatial ventriloquism for invisible trials (subjective awareness criterion):

Analysis 1 was restricted to only 11 participants that had at least 10 *visible* and 10 *invisible* trials in each condition. To confirm the robustness of the ventriloquist effect for invisible trials from analysis 1, analysis 2 focused only on invisible trials. This enabled us to include 25 participants in total that had at least 10 *invisible* trials in each condition. As in analysis 1, the perceived sound location was entered into a 2 (flash location: left, right) x 3 (sound location: left, centre, right) repeated measures ANOVA. This ANOVA identified a significant main effect of sound location (F (1.129, 27.089) = 137.082, p < 0.001) and flash location (F (1, 24) = 5.201, p = 0.032) in the absence of a significant interaction (F (1.762, 42.297) = 0.114, p = 0.869). Moreover, directly comparing the perceived sound location for left and right flashes (averaged across the three sound locations, as in post hoc analysis 1) in a directional paired t-test replicated the results of analysis 1 in a larger subset of participants and confirmed that even invisible flashes induce a significant spatial bias on sound localization (*t*(24) = 2.28, p = 0.016) (**Figure 2B**).

Analysis 3 - Spatial ventriloquism for invisible trials (objective awareness criterion):

Analysis 1 and 2 demonstrated a ventriloquist effect for flashes that participants judged invisible, i.e. that they were unaware of according to a subjective 'awareness' criterion. In analysis 3, we asked whether invisible flashes are able to influence the perceived sound location, even when participants are at chance when locating the visual flash (i.e. objective awareness criterion). This analysis was enabled by the inclusion of additional catch trials in the experiment where participants reported not only the location of the sound and the visibility of the flash but also the location of the flash. Based on these catch trials, we identified 22 participants (out of the 25 participants from analysis 2) that showed 'chance performance' when localizing a flash that was judged invisible by them. This 2 (flash location: left, right) x 3 (sound location: left, centre, right) repeated measures ANOVA basically replicated the results of analysis 2. Critically, it confirmed a significant shift in perceived auditory location induced by invisible flashes, even when participants were not better than chance to locate these invisible flashes (*t* (21) = 1.855, p = 0.039) (**Figure 2B**).

Correlation between visible and invisible ventriloquist effects:

Next we investigated whether the strength of the ventriloquist effect for visible and invisible trials is correlated over participants. A correlation may provide initial evidence that there is one underlying multisensory process mediating the ventriloquist effect for visible and invisible trials. In contrast to this conjecture, the ventriloquist effect was not correlated for visible and invisible trials over participants ($r = 0.338$, $p = 0.155$). While null-results need to be considered with great caution, this lack of correlation may suggest that different processes or circuitries may mediate audiovisual integration for visible and invisible flashes (**Figure 3**).

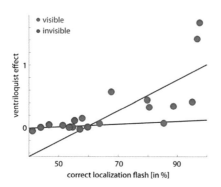

Figure 3. Scatter plots depict the regression of the ventriloquist effect against localization accuracy of the visual flash separately for visible (blue) and invisible (green) trials. The ordinate represents the ventriloquist effect averaged over all sound locations and the abscissa represents visual localization accuracy (i.e. % correct).

Influence of reliability of flash location information on spatial ventriloquism:

According to current models of multisensory integration (Ernst and Banks, 2002; Alais and Burr, 2004), the influence of the flash location on the perceived auditory location should increase with the reliability of the visual spatial information. While the experimental design did not allow us to test this prediction in a quantitative sense, in a first approximate qualitative analysis we investigated whether the ventriloquist effect depended on participants' accuracy (or precision) with which they were able to locate the flash. We predicted that at least for visible flashes, higher visual localization accuracy is associated with a greater ventriloquist effect. Indeed, for visible flashes, participants' localization performance accuracy positively predicts the size of their ventriloquist effect ($r = 0.626$, Beta = 2.422, $p = 0.053$). In other words, the ventriloquist effect increased with their localization performance. However, no significant relationship was detected between the ventriloquist effect and visual spatial localization performance, when the flash was invisible ($r = 0.301$, Beta = 0.223, $p = 0.398$). Further, the regression slope for visible trials was significantly greater than for invisible trials ($t_{(7)} = 1.952$, $p = 0.0459$). These results provide further evidence that visual representations may interact with sound processing without being accessible to perceptual awareness or visual location tasks.

DISCUSSION

Using continuous flash suppression we demonstrate that spatial ventriloquism is profoundly modulated by perceptual awareness. While a robust ventriloquist effect was observed for visual flashes that were consciously perceived, it was nearly abolished for physically identical flashes that evaded participants' awareness. Nevertheless, even flashes that participants did not consciously perceive induced a small, but significant ventriloquist effect. These results suggest that auditory and visual inputs are integrated into coherent spatial representations at both pre- and post-aware processing stages.

The strong effect of awareness on spatial ventriloquism challenges the classical views of automatic spatiotemporal integration (Driver, 1996; Bertelson and Aschersleben, 1998; Bertelson et al., 2000a; Vroomen et al., 2001; Helbig and Ernst, 2008) that have evolved from numerous psychophysics studies demonstrating that spatial ventriloquism is largely unaffected by attention (Bertelson et al., 2000a; Vroomen et al., 2001; Vroomen and de Gelder, 2004).

Yet, our results dovetail nicely with current perspectives on the neural organization of multisensory integration. Specifically, auditory and visual information are thought be integrated via multiple circuitries including subcortical mechanisms, direct connectivity between primary sensory areas and convergence in higher order association areas (Macaluso and Driver, 2005; Ghazanfar and Schroeder, 2006; Musacchia and Schroeder, 2009; Kayser et al., 2012). Moreover, it is well established that multisensory integration progressively increases along the cortical hierarchy with only about 15 % neurons showing multisensory properties in primary sensory areas (Bizley et al., 2007) and more than 50 % in classical association areas such as intraparietal or superior temporal sulci (Dahl et al., 2009).

Thus, when the visual flash escapes the continuous flash suppression and enters participants' awareness, a strong ventriloquist effect emerges most likely via integration in association areas such as IPS that contain exuberant multisensory neurons. By contrast, when the continuous flash suppression successfully blocks neural activity from propagating beyond V1, audiovisual integration in higher order association areas is greatly attenuated or even abolished leading to a smaller ventriloquist effect. Nevertheless, these 'suppressed invisible' flashes may influence sound localization via at least two distinct neural circuitries. First, an invisible flash may modulate sound processing via sparse direct connectivity between primary auditory and visual areas (Falchier et al., 2002; Cappe and Barone, 2005). Second, it may interact with auditory signals via subcortical mechanisms such as the collicular or pulvinar pathway (Wallace et al., 1993; Hackett et al., 2007; Cappe et al., 2009b; Cappe et al., 2009a) that has previously been implicated in mediating activations along the dorsal stream into the intraparietal sulcus under CFS (Fang and He, 2005). The ventriloquist effect may be smaller for invisible flashes, because these two circuitries may be less effective than classical association areas (e.g. IPS) in mediating visual influences at a representational level (i.e. spatial location) on sound processing. Alternatively, flashes that do not enter participants' awareness may elicit representations that provide less reliable spatial information than visible flashes and hence induce a smaller ventriloquist effect. According to this second account, physically identical flashes evoke spatial representations that differ in their precision already at the primary cortical level (V1) because of state-dependent effects or various sources of neural noise. While strong activations associated with reliable V1 representations are able to break through the flash suppression and propagate into IPS, less reliable activations for 'invisible flashes' are blocked from higher order cortical processing.

Evidence for less reliable spatial representations for invisible flashes is indicated by participants' visual localization performance during catch trials. Participants were far less precise in locating the position of invisible than visible flashes. However, even when participants performed at chance when locating invisible flashes, i.e. they were unaware of the flashes according to an objective criterion of awareness, a small ventriloquist effect remained. In other words, flashes that did not provide any reliable spatial information as assessed by the visual localization task nevertheless influenced participants' perceived sound location. Thus, neural spatial representations of the flash may be able to influence sound processing via neural circuitries that are not accessible for visual localization and perceptual awareness.

Further evidence for potentially distinct neural processes underlying audiovisual integration in the presence and absence of awareness is provided by the additional regression analyses over participants that investigated whether flash localization performance predicted the ventriloquist effect. Based on current computational models of multisensory integration, one would expect that the ventriloquist effect increases with the accuracy of participants' visual localization performance (Alais and Burr, 2004). Indeed, when the flash was visible, the ventriloquist effect was positively predicted by the accuracy with which participants were able to locate the flash. However, when the flash was invisible no significant relationship was observed. These differences support the idea that visible and invisible flashes may influence sound processing potentially via distinct processes and neural circuitries. Future studies using EEG and fMRI will be able to dissociate the neural circuitries that enable visual stimuli to influence sound localization in the presence and absence of awareness.

MATERIALS AND METHODS

2.1 Participants

After giving informed consent, 32 healthy young adults (19 females, 30 right-handed, mean age: 23.5 years, standard deviation: 3.53, range: 18-38) with normal or corrected-to-normal vision, participated in this study. No subject reported any hearing deficits. The study was approved by the local ethics review board of the University of Tübingen.

2.2 Stimuli and apparatus

Participants sat in a dimly lit room in front of a computer monitor at a viewing distance of 1 m. They viewed one half of the monitor with each eye using a custom-built mirror stereoscope. Visual stimuli were composed of targets and masks that were presented on a grey, uniform background with a mean luminance of 15.5 cd/m^2. One eye viewed the target stimuli, i.e. two grey discs (Ø 0.29°, mean luminance: 25.4 cd/m^2), located 5.72° visual angle to the left and right of a grey fixation dot. On each trial, either the left or the right target's colour changed to white (mean luminance: 224.2 cd/m^2) for a duration of 100 ms. This change in brightness will be referred to as 'flash'. To suppress the flash's perceptual visibility, two dynamic Mondrians (Ø 2°) were shown to the other eye (Maruya et al., 2008). To match the target's location the Mondrians' were also centred 5.72° to the left and right of the fixation dot. Each Mondrian consisted of sinusoidal gratings (Ø 0.57°) which changed their colour and position randomly at a frequency of 10 Hz. Each grating's texture was shifted every 16.6 ms to generate apparent motion. Visual stimuli were presented foveally, contained a fixation spot and were framed by a grey, isoluminant square aperture of 8.58° x 13.69° in diameter to aid binocular fusion.

Auditory stimuli were pure tones with a carrier frequency of 1 kHz and a duration of 100 ms. They were presented via six external speakers, placed above and below the monitor. Upper and lower speakers were aligned vertically and located centrally, 2.3° to the left and 2.3° to the right of the monitor's centre. Speakers' location was chosen by trading off physical alignment of visual and auditory stimulus locations and sound localization performance. At a distance of 2.3° mean sound localization accuracy amounted to ~70% (see below).

Psychophysical stimuli were generated and presented on a PC running Windows XP using the Psychtoolbox version 3 (Brainard, 1997; Kleiner et al., 2007) running on Matlab 7 (Mathworks, Nantucket, Massachusetts). Visual stimuli were presented dichoptically using a gamma-corrected 30″ LCD monitor with a resolution of 2560 x 1600 pixels at a frame rate of 60Hz (GeForce 8600GT graphics card). Auditory stimuli were digitized at a sampling rate of 44.8 kHz via a M-Audio Delta 1010LT sound card and presented at an maximal amplitude of 73 dB sound pressure level. Exact audiovisual onset timing was confirmed by recording visual and auditory signals concurrently with a photo-diode and a microphone.

2.3 Experimental Design

In a ventriloquist paradigm, the 2 x 3 x 2 design manipulated (1) 'flash location' (2 levels: left flash and right flash) and (2) 'sound location' (3 levels: left sound, central sound and right sound) and (3) flash visibility (visible, invisible). (**Figure 1A**).

Each trial started with the presentation of the fixation dot for a duration of 1000 ms. Next, participants' one eye was presented with two grey discs, located 5.72° visual angle to the left and right of a grey fixation dot. Participants' awareness of these discs was suppressed by showing dynamic Mondrians to the other eye (i.e. dynamic flash suppression). The assignment of eyes was changed after each trial, to enhance suppression. After a random interval of 500-1000 ms either the left or the right disc 'flashed', i.e. changed its luminance for a duration of 100 ms. In synchrony with the flash, an auditory beep was played from one of three potential locations. On each trial, participants reported the location of the beep (left, centre, right) and rated the visibility of the flash (visible, unsure, invisible). The 'unsure' response option was primarily included to encourage participants to categorize trials as invisible when they were fully unaware of the flash, but not of further interest for the analysis. In addition, on 22.2% of the trials, the so-called catch trials, participants were also asked to locate the flash. The Mondrian masks and the discs were presented on the screen until the participants responded to all questions. Participants responded by pressing one of three buttons on a keyboard. The button assignment was counterbalanced across participants. For an example trial see **figure 1B**.

This visibility judgment provided a 'subjective awareness criterion'. Critically, the flash was visible only in a fraction of trials allowing us to quantify the effect of awareness on multisensory integration by comparing spatial ventriloquism for visible and invisible - yet physically identical - flashes. The catch trials with the additional flash localization allowed us to assess the reliability of the spatial information about the flash provided by visible and invisible trials. Furthermore, based on binomial tests, we were able to select participants that were significantly better than chance when locating visible flashes (i.e. 28 'reliable participants' included in the study) and not better than chance when locating invisible flashes (i.e. objective awareness criterion).

Prior to the main experiment, participants were familiarized with stimuli and task. First, they completed 2-3 sessions of sound localization (% correct day1: 69.9% (std.: 16.8); % correct day2: 74.7% (std.: 14.7), % correct both days: 72.3% (std.14.9)). Next, there were two short practice sessions of the ventriloquist paradigm. During the main experiment participants completed a total of 24 experimental sessions distributed over two successive days, resulting in a total of 1296 trials (i.e. 216 trials per condition).

2.4 Analysis

Out of the 32 participants, only 28 participants that localized the flash significantly better than chance were included in this study (i.e. based on binomial tests on flash localization judgments during the catch trials).

For each of those participants, we estimated the *perceived auditory location* for each condition and visibility level by averaging their localization responses that were initially coded as -1 for left, 0 for centre and 1 for right across trials. This perceived location was then used as the dependent variable for the following analyses:

Analysis 1 – Effect of subjective visibility on spatial ventriloquism:

In analysis 1, we investigated whether spatial ventriloquism is modulated by perceptual awareness of the flash in a 2 (flash location: left, right) x 3 (sound location: left, right) x 2 (flash visibility: visible, invisible) repeated measures ANOVA on perceived sound location. To ensure reliable parameter estimation, this analysis included only the 11 participants that had at least 10 trials per condition and visibility level.

Analysis 2 – Spatial ventriloquism for invisible trials (subjective awareness criterion):

In analysis 2, we confirmed the robustness of the ventriloquist effect for invisible trials, in a 2 (flash location: left, right) x 3 (sound location: left, centre, right) repeated measures ANOVA on perceived sound location. Focusing only on invisible trials, we were able to include 25 participants in total that had at least 10 *invisible* trials in each condition.

Analysis 3 - Spatial ventriloquism for invisible trials (objective awareness criterion):

In analysis 3, we asked whether invisible flashes were able to influence the perceived sound location, even when participants were at chance when locating the visual flash (i.e. objective awareness criterion). Based on the catch trials, we identified 22 participants (out of the 25 participants from analysis 2) that showed 'chance performance' when localizing a flash that was judged invisible by them (based on the binomial distribution).

Correlation between visible and invisible ventriloquist effects:

We investigated whether the strength of the ventriloquist effect for visible and invisible trials (i.e. limited to the 11 participants of analysis 1) are correlated over participants using Pearson correlations.

Influence of reliability of flash location information on spatial ventriloquism:

We investigated whether the ventriloquist effect depends on participants' accuracy (or precision) with which they were able to locate the flash (n.b. this analysis included 10 participants from analysis 1, one subject had to be excluded as there was only one invisible catch trial). For this, we performed two regression analyses where we used (i) the flash localization accuracy during invisible trials to predict the ventriloquist effect during invisible trials and (ii) the flash localization accuracy during visible trials to predict the ventriloquist effect during visible trials. Finally, we investigated whether the regression slopes were significantly different in a regression model where we used the difference in flash localization accuracy between visible and invisible trials to predict the difference in ventriloquist between visible and invisible trials.

REFERENCES

Alais D, Burr D (2004) The ventriloquist effect results from near-optimal bimodal integration. Curr Biol 14:257-262.

Alsius A, Navarra J, Campbell R, Soto-Faraco S (2005) Audiovisual integration of speech falters under high attention demands. Curr Biol 15:839-843.

Bertelson P, Radeau M (1981) Cross-modal bias and perceptual fusion with auditory-visual spatial discordance. Percept Psychophys 29:578-584.

Bertelson P, Aschersleben G (1998) Automatic visual bias of perceived auditory location. Psychonomic Bulletin & Review 5:482-489.

Bertelson P, Vroomen J, de Gelder B, Driver J (2000a) The ventriloquist effect does not depend on the direction of deliberate visual attention. Percept Psychophys 62:321-332.

Bertelson P, Pavani F, Ladavas E, Vroomen J, de Gelder B (2000b) Ventriloquism in patients with unilateral visual neglect. Neuropsychologia 38:1634-1642.

Bizley JK, Nodal FR, Bajo VM, Nelken I, King AJ (2007) Physiological and anatomical evidence for multisensory interactions in auditory cortex. Cereb Cortex 17:2172-2189.

Brainard DH (1997) The Psychophysics Toolbox. Spat Vis 10:433-436.

Cappe C, Barone P (2005) Heteromodal connections supporting multisensory integration at low levels of cortical processing in the monkey. Eur J Neurosci 22:2886-2902.

Cappe C, Rouiller EM, Barone P (2009a) Multisensory anatomical pathways. Hear Res 258:28-36.

Cappe C, Morel A, Barone P, Rouiller EM (2009b) The thalamocortical projection systems in primate: an anatomical support for multisensory and sensorimotor interplay. Cereb Cortex 19:2025-2037.

Dahl CD, Logothetis NK, Kayser C (2009) Spatial organization of multisensory responses in temporal association cortex. J Neurosci 29:11924-11932.

Driver J (1996) Enhancement of selective listening by illusory mislocation of speech sounds due to lip-reading. Nature 381:66-68.

Ernst MO, Banks MS (2002) Humans integrate visual and haptic information in a statistically optimal fashion. Nature 415:429-433.

Falchier A, Clavagnier S, Barone P, Kennedy H (2002) Anatomical evidence of multimodal integration in primate striate cortex. J Neurosci 22:5749-5759.

Fang F, He S (2005) Cortical responses to invisible objects in the human dorsal and ventral pathways. Nat Neurosci 8:1380-1385.

Ghazanfar AA, Schroeder CE (2006) Is neocortex essentially multisensory? Trends Cogn Sci 10:278-285.

Hackett TA, De La Mothe LA, Ulbert I, Karmos G, Smiley J, Schroeder CE (2007) Multisensory convergence in auditory cortex, II. Thalamocortical connections of the caudal superior temporal plane. J Comp Neurol 502:924-952.

Helbig HB, Ernst MO (2008) Visual-haptic cue weighting is independent of modality-specific attention. J Vis 8:21 21-16.

Kayser C, Petkov CI, Remedios R, Logothetis NK (2012) Multisensory Influences on Auditory Processing: Perspectives from fMRI and Electrophysiology. In: The Neural Bases of Multisensory Processes (Murray MM, Wallace MT, eds). Boca Raton (FL).

Kleiner M, Brainard D, Pelli D (2007) What's new in Psychtoolbox-3? Perception 36.

Macaluso E, Driver J (2005) Multisensory spatial interactions: a window onto functional integration in the human brain. Trends Neurosci 28:264-271.

Maruya K, Watanabe H, Watanabe M (2008) Adaptation to invisible motion results in low-level but not high-level aftereffects. J Vis 8:7 1-11.

Munhall KG, ten Hove MW, Brammer M, Pare M (2009) Audiovisual integration of speech in a bistable illusion. Curr Biol 19:735-739.

Musacchia G, Schroeder CE (2009) Neuronal mechanisms, response dynamics and perceptual functions of multisensory interactions in auditory cortex. Hear Res 258:72-79.

Radeau M, Bertelson P (1977) Adaptation to auditory-visual discordance and ventriloquism in semirealistic situations. Attention, Perception, & Psychophysics 22:137-146.

Talsma D, Senkowski D, Soto-Faraco S, Woldorff MG (2010) The multifaceted interplay between attention and multisensory integration. Trends Cogn Sci 14:400-410.

Tsuchiya N, Koch C (2005) Continuous flash suppression reduces negative afterimages. Nat Neurosci 8:1096-1101.

Vroomen J, de Gelder B (2004) Perceptual effects of cross-modal stimulation: Ventriloquism and the freezing phenomenon. In: The handbook of multisensory processes (Calvert G, Spence C, Stein BE, eds), pp 141 - 146. MA, USA: MIT Press.

Vroomen J, Bertelson P, de Gelder B (2001) The ventriloquist effect does not depend on the direction of automatic visual attention. Percept Psychophys 63:651-659.

Wallace MT, Meredith MA, Stein BE (1993) Converging influences from visual, auditory, and somatosensory cortices onto output neurons of the superior colliculus. J Neurophysiol 69:1797-1809.

Watanabe M, Cheng K, Murayama Y, Ueno K, Asamizuya T, Tanaka K, Logothetis N (2011) Attention but not awareness modulates the BOLD signal in the human V1 during binocular suppression. Science 334:829-831.

CHAPTER 2:

STEADY-STATE RESPONSES DEMONSTRATE INFORMATION INTEGRATION WITHIN BUT NOT ACROSS THE SENSES

Anette S. Giani[1], Erick B. Ortiz[2], Paolo Belardinelli[2], Mario Kleiner[1], Hubert Preissl[2,3], Uta Noppeney[1]

[1]Max Planck Institute for Biological Cybernetics, Tübingen, Germany
[2]MEG Centre, University of Tübingen, Tübingen, Germany
[3]Department of Obstetrics and Gynecology, University of Arkansas for Medical Sciences, Little Rock, USA

ABSTRACT

To form a unified percept of our environment, the human brain integrates information within and across the senses. This MEG study investigated interactions within and between sensory modalities using a frequency analysis of steady-state responses that are elicited time-locked to periodically modulated stimuli. Critically, in the frequency domain, interactions between sensory signals are indexed by crossmodulation terms (i.e. the sums and differences of the fundamental frequencies). The 3×2 factorial design, manipulated (1) modality: auditory, visual or audiovisual (2) steady-state modulation: the auditory and visual signals were modulated only in one sensory feature (e.g. visual gratings modulated in luminance at 6 Hz) or in two features (e.g. tones modulated in frequency at 40 Hz & amplitude at 0.2 Hz). This design enabled us to investigate crossmodulation frequencies that are elicited when two stimulus features are modulated concurrently (i) in one sensory modality or (ii) in auditory and visual modalities. In support of within-modality integration, we reliably identified crossmodulation frequencies when two stimulus features in one sensory modality were modulated at different frequencies. In contrast, no crossmodulation frequencies were identified when information needed to be combined from auditory and visual modalities. The absence of audiovisual crossmodulation frequencies suggests that the previously reported audiovisual interactions in primary sensory areas may mediate low level spatiotemporal coincidence detection that is prominent for stimulus transients but less relevant for sustained SSR responses. In conclusion, our results indicate that information in SSRs is integrated over multiple time scales within but not across sensory modalities at the primary cortical level.

INTRODUCTION

How does the human brain integrate information within and across sensory modalities to form a unified percept? This question has traditionally been addressed using transient stimuli, analysed in the time domain. Previous functional imaging studies have demonstrated integration of different types of visual information (e.g. form, motion or colour) in visual areas at multiple levels of the cortical hierarchy (Sarkheil et al., 2008; Seymour et al., 2010). Likewise, integration of information from multiple senses emerges in a widespread system encompassing subcortical, primary sensory and higher order association cortices (Wallace et al., 1996; Calvert, 2001; Schroeder and Foxe, 2002; Beauchamp et al., 2004; van Atteveldt et al., 2004; Kayser et al., 2007; Lakatos et al., 2007; Ghazanfar et al., 2008; Noesselt et al., 2008; Sadaghiani et al., 2009; Noppeney et al., 2010; Werner and Noppeney, 2010b, a). Magnetoencephalography and electroencephalography (M/EEG) demonstrated that multisensory integration emerges very early at around 50–100 ms after stimulus onset (Schroger and Widmann, 1998; Giard and Peronnet, 1999; Fort et al., 2002; Molholm et al., 2002; Talsma and Woldorff, 2005; Teder-Salejarvi et al., 2005; Talsma et al., 2007; Raij et al., 2010).

Identification of information integration and -interplay in the time and frequency domains - methodological considerations

In the time domain, integration of transient sensory stimuli is commonly determined by comparing the response to transient audiovisual (AV) stimuli to the summed responses for auditory (A) and visual (V) stimuli (i.e. AV≠A+V). The rationale of this so-called super- or subadditivity criterion is that under the null-hypothesis of no audiovisual integration or interaction, the response to the audiovisual compound stimulus should be a linear combination (i.e. sum) of the responses to the two unisensory stimulus components when presented alone. However, despite its underlying rationale, this approach is often confounded by non-specific stimulus evoked processes such as arousal, anticipatory processes etc. (Teder-Salejarvi et al., 2002; Goebel and van Atteveldt, 2009). Since these general cognitive processes are elicited by each stimulus irrespective of its uni- or multisensory nature, they are counted twice for the sum of the two unisensory responses but only once for the bisensory response rendering the comparison unbalanced (Teder-Salejarvi et al., 2002; Noppeney, 2011).

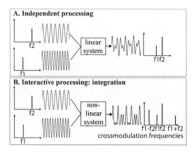

Figure 1. Responses to two sinusoidal signals in the time and frequency domain: **(A)** Independent processing: If two sinusoidal inputs at frequencies f1 and f2 are processed independently (= linear system), the output is a linear combination (e.g. sum) of the two component signals (with potential changes in phase and amplitude). Hence, in the frequency domain, power is observed only at frequencies f1 and f2. **(B)** Interactive processing (= integration): If two sinusoidal inputs at frequencies f1 and f2 are processed interactively (= non-linear system), the output signal represents the multiplication of the two input signals. Hence, in the frequency domain, power is observed also at the crossmodulation frequencies, i.e. the sums and differences of the input frequencies (f1±f2).

Alternatively, integration of sensory signals within and across the senses can be studied in the frequency domain using steady-state responses (SSRs). The term 'SSR' refers to oscillatory brain activity that is elicited by and time-locked to periodically modulated sensory stimuli, such as amplitude modulated tones or luminance modulated visual stimuli (for a review on auditory SSRs see: (Picton et al., 2003); for a review on visual SSRs see: (Vialatte et al., 2010)). Two sensory stimuli, modulated at frequencies f1 and f2, will evoke SSRs at fundamental and harmonic frequencies (i.e., n*f1 or m*f2, with m and n being any integer values). Critically, when perturbed concurrently with two periodically modulated sensory signals at two different frequencies, the brain can process and respond to them in two distinct ways: First, it may treat the two signals independently (Fig. 1A). In the time domain, we would then observe a linear combination (e.g. sum) of the two steady state responses that are individually elicited when the brain is presented with one signal alone. A frequency analysis of these steady state responses will then just reveal power at the fundamental and harmonic frequencies of the two component signals. Second, the brain may integrate the two sensory signals (Fig. 1B). In the time domain, integration is characterized by an interaction between the two sensory signals (i.e. a multiplication of the time courses of the two component signals). Importantly, interactions of two sensory signals in the time-domain are expressed in terms of power at the crossmodulation frequencies (i.e. n*f1±m*f2) in the frequency domain (Regan and Regan, 1988b, a). Hence, crossmodulation frequencies provide an alternative way to identify integration of sensory signals within the human brain. For high frequency steady-state signals, multisensory integration can be much more clearly identified in the frequency- than in the time domain.

In addition to this 'true' signal integration, two signals may influence each other's processing in a non-specific fashion in the time-domain; for instance, one signal may increase the saliency of another signal via stimulus evoked arousal and related attentional mechanisms. These non-specific modulatory effects are reflected in an amplification (or suppression) of the amplitude of the fundamental and harmonic frequencies in the frequency domain and will not induce any crossmodulation frequency terms.

EEG and MEG evidence for information integration within and across the senses using SSRs

Over the past decade, SSR paradigms have accumulated evidence for signal integration within a single sensory modality on the basis of crossmodulation frequencies. Within the auditory modality, several MEG and EEG studies have demonstrated the emergence of crossmodulation frequencies when the auditory stimuli were amplitude modulated at two frequencies or simultaneously modulated in amplitude and frequency suggesting that information about a signal's amplitude and frequency interact along the auditory processing stream (Dimitrijevic et al., 2001; Draganova et al., 2002; Luo et al., 2006; Ding and Simon, 2009). Likewise, within the visual modality, crossmodulation frequencies indicated interactions of brain signals induced by multiple visual objects that were flickering at multiple different frequencies (Ratliff and Zemon, 1982; Regan and Regan, 1988b; Fuchs et al., 2008; Sutoyo and Srinivasan, 2009).

In contrast, there is only sparse evidence for non-linear interactions across modalities. In fact, to our knowledge only one study has identified interactions between auditory and visual steady-state signals based on audiovisual crossmodulation frequencies (Regan et al., 1995). Yet, this very early study did not report statistics, effect sizes or number of participants limiting the conclusions that can be drawn from it. Hence, a new more thorough study is needed to investigate whether crossmodulation frequencies emerge when the brain is perturbed with two steady state signals in two sensory modalities.

Relevance of SSRs in cognitive neuroscience

The characterization and identification of crossmodulation frequencies as a useful index for multisensory integration would open new research avenues for tracking the influence cognitive processes in multisensory integration. Thus, since SSRs and their crossmodulation frequencies are determined by the periodicity of the stimulating signal, they can be used to 'tag and track' multiple brain signals simultaneously. For instance, frequency tagging has been used to investigate the influence of awareness of several simultaneous signals during binocular rivalry. Presenting two distinct images that were flashed at two different frequencies to the two eyes demonstrated that the SSRs' magnitude elicited by each visual stimulus is modulated by its perceptual dominance (Tononi et al., 1998; Cosmelli et al., 2004; Srinivasan, 2004; Kamphuisen et al., 2008; Sutoyo and Srinivasan, 2009). Along those lines, crossmodulation frequencies would then offer a unique opportunity to study the effect of consciousness or attention on integration of specific signals or information within and across the senses non-invasively within the human brain. For instance, one could easily and very precisely address the question whether visual signals that the subject is not aware of can interact with auditory signals that the subject is aware of.

Experimental design and hypotheses

Previous studies have demonstrated that cue integration and interplay differs when the sensory cues are from the same or from different modalities (Duncan et al., 1997; Hillis et al., 2002). Most prominently, Hillis et al. (2002) demonstrated that single cue information is lost only for integration within but not across the senses.

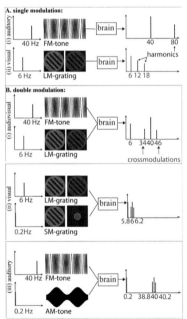

Figure 2. Experimental predictions for SSRs in the frequency domain: **(A)** Stimulation conditions with one sinusoidal signal: We expect SSRs at fundamental and harmonic frequencies: (i) auditory: n*40 Hz (with n being any integer value) for a 40 Hz frequency modulated tone (FM-tone) and (ii) visual: n*6 Hz for a 6 Hz luminance modulated grating. **(B)** Stimulation conditions with two (or more) sinusoidal signals: If the two concurrent sensory signals are processed interactively (= integration) by the brain, we expect SSRs at fundamental, harmonic and cross-modulation frequencies: (i) audiovisual: n*40±m*6 Hz for 40 Hz FM-tones and 6 Hz LM-gratings, (ii) visual: n*6±m* 0.2 Hz for a visual grating that is modulated in luminance (LM) at 6 Hz and in size (SM) at 0.2 Hz, (iii) auditory: n*40±m* 0.2 Hz for a tone that is modulated in frequency (FM) at 6 Hz and in amplitude (AM) at 0.2 Hz.

49

This MEG study was designed to investigate information integration within and across the senses. Therefore, we employed a 3×2 factorial design, manipulating (1) modality: auditory, visual or audiovisual (2) temporal dynamics: single or double modulated. In the single modulated conditions, subjects were presented with an auditory signal modulated in frequency at 40 Hz and/or with a visual grating modulated in luminance at 6 Hz. In the double modulated conditions, the visual grating was additionally modulated at 0.2 Hz in size and the auditory tone at 0.2 (or 0.7) Hz in amplitude. This design enabled us to investigate crossmodulation frequencies that are elicited when two stimulus features are modulated concurrently (i) in one sensory modality or (ii) in auditory and visual modalities. More specifically, integration within the senses was identified by testing for crossmodulation terms between e.g. 0.2 Hz and 6 Hz in the double modulated visual conditions (and likewise in the auditory conditions). Conversely, integration across the senses was identified by testing for crossmodulation terms between 6 Hz and 40 Hz, i.e. the fundamental frequencies of the visual and auditory SSRs (Fig. 2).

In the next step, we then investigated whether integration across the senses depends on the synchrony of the slow modulations of the visual and auditory signals, i.e. whether the slow modulations were identical (auditory and visual both at 0.2) or different (auditory at 0.7 and visual at 0.2) in the two sensory modalities. Hence, the double modulation conditions enabled us to ask simultaneously two questions: (i) whether signals are combined within the senses and (ii) whether additional synchrony cues provided by other stimulus features enable or enforce integration across the senses. Moreover, investigating information integration within and across the senses within the same experiment is critical for the interpretation of potential null-results that may occur in the case of the less well-established integration across the senses. A replication of crossmodulation terms for integration within the senses demonstrates that our acquisition and analysis methods are adequate for identifying crossmodulation terms and thereby provide interpretational validity for a potential absence of crossmodulation terms in across-sensory integration.

To conclude, this paradigm allowed us to study signal integration simultaneously within and across modalities. Moreover, we investigated the role of synchrony, which has been shown to be a major determinant of multisensory integration both at the neuronal and behavioural levels (Wallace et al., 2004; Noesselt et al., 2008; Lewis and Noppeney, 2010)

MATERIALS AND METHODS

Participants

After giving informed consent, fourteen healthy young adults (5 females, 12 right-handed, mean age: 24.8 years, standard deviation: 3.3, range: 20–30) with normal or corrected-to-normal vision, participated in this study. No subject reported any hearing deficits. Due to excessive non-physiological artefacts, two additional female participants were excluded from the analysis. The study was approved by the local ethics review board of the University of Tübingen.

Stimuli

The visual stimulus was a red square-wave grating delimited by a red circle. To evoke visual SSRs, the gratings' luminance was modulated sinusoidally at a rate of 6 Hz and a modulation depth of 100%. During single modulation (SM) conditions, the grating's size (i.e. the grating's overall size) was held constant (7.6° visual degrees). During double modulation (DM) conditions, the grating's size was additionally modulated sinusoidally at a frequency of 0.2 Hz (3.1°–12.2° visual degrees) (Fig. 3B, rows 1–2). The grating together with a light red fixation dot was presented in the centre of the screen on a black background.

The auditory stimulus was a pure tone with a carrier frequency of 500 Hz. To evoke an auditory SSR, the tone's frequency was modulated sinusoidally at a rate of 40 Hz with a maximum frequency deviation of 350 Hz. During SM conditions, the tone's amplitude was held constant. During DM conditions, the tone's amplitude was additionally modulated at a frequency of 0.2 Hz or 0.7 Hz with a modulation depth of 60% (Fig. 3B, rows 3–5). The sound's overall amplitude was adjusted such that the root-mean-square of the sound vector was held constant over trials.

For the single modulation conditions, the modulation frequencies were selected based on pilot studies and previous research, showing that SSRs are most effectively evoked in the auditory domain by pure tones with a carrier frequency of 500 Hz and modulation rates at ~40 Hz (Galambos et al., 1981; Hari et al., 1989; Ross et al., 2000) and in the visual domain by modulation frequencies ranging from 5 to 20 Hz.

For the double modulation conditions, the additional modulation frequencies were selected <1 Hz in order to enable the perception of audiovisual synchrony.

Both, auditory and visual stimuli were generated and presented using Matlab, Psychtoolbox version 3 (revision 1754) (Brainard, 1997; Kleiner et al., 2007) (http://psychtoolbox.org) running on an Apple MacBook Pro with Nvidia GeForce 9440 M graphics card under Macintosh OS-X 10.6.7. A 60 Hz LCD projector (Sanyo, PLC-XP41), located outside the magnetically shielded room, was used to project the visual stimulus onto a focusing screen, located approximately 80 cm in front of the subject's head. Tones were digitized at a sampling rate of 44.8 kHz via the computer's internal HDA sound chip and presented binaurally via insert earphones (E-A-RTONE® 3A, Aero Company, USA) at a maximal amplitude of 75 dB sound pressure level (SPL).

Experimental design and procedure

The experiment conformed to a 3×2 design manipulating (1) 'stimulation modality' (3 levels): A (auditory only), V (visual only) and AV (audiovisual) and (2) 'steady-state modulation' (2 levels): single modulation (SM), double modulation (DM1 or DM2). (Fig. 3A). During single modulation conditions, only one stimulus feature was modulated per modality. More specifically, the grating's luminance and/or the tone's frequency were modulated at frequencies 6 Hz and 40 Hz, respectively. During double modulation conditions, additional slow modulations were imposed on the auditory and visual steady-state stimuli. For auditory stimulation, we additionally modulated the tone's amplitude at a rate of 0.2 Hz during DM1 and 0.7 Hz during DM2 conditions. For visual stimulation, we additionally modulated the grating's size at a rate of 0.2 Hz for both DM1 and DM2 conditions (n.b. since for visual presentation DM-V1=DM-V2, they are referred to as DM-V trials). During double modulated audiovisual conditions (i.e. DM-AV), the slow auditory and visual modulation frequencies were either identical (i.e. DMAV1= synchronous) or different (DM-AV2=asynchronous). We also included a fixation condition, in which only the fixation dot was presented. The experimental design thus encompassed 8 activation conditions and 1 fixation condition, resulting in a total of 9 conditions (Fig. 3).

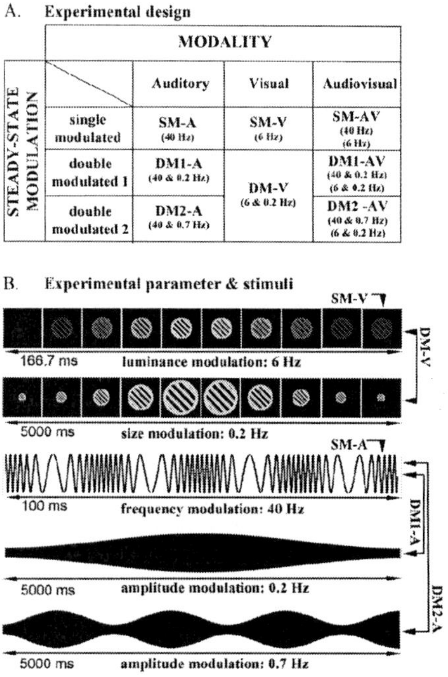

A. **Experimental design**

B. **Experimental parameter & stimuli**

Figure 3. Experimental design and stimuli. **(A)** The 3 × 2 factorial design manipulates: 1) modality: auditory only (A), visual only (V) and audiovisual (AV); 2) steady-state modulation: single modulated (SM) and double modulated (DM1 and DM 2). DM1-V and DM2-V conditions are identical and are therefore referred to as DM-V. **(B)** Experimental parameters and Stimuli: Single modulated (SM) stimuli are luminance modulated gratings (visual; row1) and frequency modulated tones (auditory; row 3). Double modulated stimuli (DM) are (i) gratings modulated concurrently in luminance and size (visual; rows 1–2) and (ii) tones modulated concurrently frequency and amplitude (auditory; rows 3–5).

Participants passively perceived the auditory and/or visual stimuli, while fixating the dot in the centre of the screen. They were instructed not to move and encouraged to blink between trials. To minimize movement artefacts, participants were comfortably lying in supine position throughout the whole experiment.

All participants completed a total of 8 sessions. Each session included 9 trials, i.e. one continuous 62 second trial per condition, with a brief inter-trial interval of 5 s. The order of conditions was counterbalanced across session and subjects.

Exploratory pilots to evaluate the effect of modulation frequency

Since SSRs at fundamental, harmonic and within-sensory crossmodulation frequencies were identified reliably at the single subject level, we investigated in an exploratory fashion whether audiovisual crossmodulation frequencies could be obtained for other parameter settings. First, we acquired single subject data for the following modulation frequencies: 3 Hz (V) and 3.5 Hz (A), 5 Hz (V) and 4 Hz (A), 8.57 Hz (V) and 19 Hz (A), 15 Hz (V) and 20 Hz (A), 30 Hz (V) and 40 Hz (A), 40 Hz (A) and 0.2 Hz (V) or 6 Hz (V) and 0.2 Hz (A). Specifically, we investigated whether crossmodulation frequencies may emerge when auditory and visual modulation frequencies are quite similar. Similar stimulation frequencies in the auditory and visual modalities are known to induce the flicker-flutter phenomenon where the modulation frequency of a sound alters subjects' perceived frequency of concurrent visual flicker (Regan and Spekreijse, 1977; Noesselt et al., 2008). Second, we investigated whether crossmodulation frequencies depend on subjects' cognitive state. For instance, it has previously been argued that audiovisual interactions of ERPs depend on subjects attending to both sensory modalities (Talsma et al., 2007). Hence, we engaged 8 subjects in a target detection task that required them to actively attend to auditory and visual modalities (Saupe et al., 2009a; Saupe et al., 2009b; de Jong et al., 2010; Gander et al., 2010). Third, since attentional effects for steady-state responses were most pronounced for brief trials (Gander et al., 2010) we subsequently shortened the trials from 60 to 4 s for 1 subject.

Data acquisition

Neuromagnetic data were recorded at 1171.88 Hz sampling frequency with a 275-channel whole-head MEG System (VSM, MedTech, Port Coquitlam, Canada; 275 axial gradiometers with 5 cm baseline and 29 reference channels) at the MEG Centre Tübingen, Germany. Subjects' head position was continuously monitored by three sensor coils attached to the nasion, and left and right pre-auricular (10 mm anterior to the left and right tragus) points of each subject. The positions of these coils, i.e. the fiducial points, were marked on the subject's skin (see below). Eye movements and blinks were recorded by a horizontal and vertical electrooculogram (EOG, for all except for 2 subjects because of their large head size).

A 3T Siemens Magnetom Tim Trio System (Siemens, Erlangen, Germany) at the MPI for Biological Cybernetics, Tübingen, Germany, was used to acquire high-resolution structural images (176 sagittal slices, TR = 2300 ms, TE = 2.98 ms, TI = 1100 ms, flip angle = 9°, FOV = 256 mm × 240 mm × 176 mm, voxel size = 1 mm × 1 mm × 1 mm). MR markers that can be identified on the anatomical image were attached to the fiducial points to enable accurate co-registration of the anatomical MRI and the MEG data.

Data analysis

The MEG and MRI data were processed and analysed using the statistical parametric mapping SPM8 (http://www.fil.ion.ucl.ac.uk/ spm/; Wellcome Trust Centre of Neuroimaging, London, UK) and Matlab 7 (Mathworks, Inc., Massachusetts, USA).

Sensor space analysis and statistics

For sensor space analysis, the continuous data per trial were notch (48–52 Hz) and bandpass (1–90 Hz) filtered in forwards and reverse directions using a 5th order Butterworth digital filter. The filtered data were downsampled to 409.6 Hz and epoched into 10 second segments (i.e. providing a frequency resolution of 0.1 Hz). Since each session encompassed one 62 second continuous trial per condition, this resulted in 6 epochs per condition and session with the first and last second of each trial being discarded to allow for SSR stabilization. The downsampling to 409.6 Hz ensured that each epoch comprised an integer number of stimulus periods for all frequencies of interest (FOIs) (Table 1) and the number of samples per epoch equalled 2^{12}.

Epochs were rejected when the MEG signal exceeded 2 pT or the EOG peak-to-peak distance exceeded 50 µV leaving on average 40.8 epochs per condition and subject. Channels were defined as bad when they contained more than 20% artefactual trials (number of bad channels ranged from 0 to 7 per subject (mean: 0.6)). After averaging the artefact-free epochs for each condition, the fast Fourier transform (FFT) was computed separately for each channel. The linearly interpolated topography × frequency data were converted to 3 D images (voxel size: 2.1 mm×2.7 mm×0.1 Hz, image dimension: 64×64×2049). The images were smoothed in space, but not in frequency, using a Gaussian Kernel of 30 mm full-width at half maximum.

1) At the random effects or between-subject level, for each subject one topography × frequency image per condition was entered in a repeated measures ANOVA modeling each of the 9 conditions in our experimental design and the subject-specific effects. Using a conjunction null conjunction analysis (i.e. a logical AND), we investigated whether the power at the fundamental and harmonic frequencies of the visual (or auditory) SSRs is increased for all visual (or auditory) conditions relative to fixation (Friston et al., 2005; Nichols et al., 2005).

2) Within- and across sensory signal integration and effects of synchrony were investigated at the relevant crossmodulation frequencies (i.e. 'true' signal integration) and at the fundamental and harmonic frequencies (i.e. non-specific crossmodal effects of saliency etc.) using paired t-tests. Specifically, we performed the following statistical comparisons:

   a) To test for effects of within-sensory signal integration, we compared the amplitudes for DM trials and SM trials separately for visual and auditory unisensory conditions.

b) To test for effects of multisensory signal integration, we compared the amplitudes of unisensory visual (or auditory) and audiovisual trials separately for each level of the factor 'steady-state modulation' (i.e. conditions: SM, DM1 and DM2).

c) To investigate the effect of synchrony, we compared the amplitudes for synchronous and asynchronous audiovisual trials (i.e. DM1-AV and DM2-AV).

Search volume constraints (i.e. frequencies of interest)

Frequencies of interest (FOIs) included fundamental frequencies and first harmonics of single modulations (i.e. visual: 6 and 12 Hz; auditory: 40 and 80 Hz) (Fig. 2A). Fundamental and harmonic frequencies of slow modulations (i.e. 0.2 and 0.7 Hz) were not analysed, because activity below 1 Hz can be caused by many non-specific external sources.

Across-sensory integration was tested at the sums and differences of the auditory and visual fundamental frequencies (40 ± 6 Hz=34 and 46 Hz) and at the sums and differences of the fundamental auditory and first harmonic visual frequency ($40\pm2*6$ Hz=28 and 52 Hz; Fig. 2B). Likewise, within-sensory integration was tested at the sums and differences of the two modulation frequencies in the double modulation conditions, i.e. for double-modulated visual at 6 ± 0.2=5.8 and 6.2 Hz (DM-V), for double-modulated auditory at 40 ± 0.2=39.8 and 40.2 Hz (DM1-A) and 40 ± 0.7=93.3 and 40.7 Hz (DM2-A) (Fig. 2B).

For a detailed overview of all statistical tests and corresponding frequencies of interest (FOIs) see Table 1. Unless otherwise stated, we report effects at $p < 0.05$ at the peak level corrected for multiple comparisons within the entire interpolated scalp space and limited to the relevant FOIs of each contrast.

To ensure not to miss any crossmodulation terms that may be expressed variably across subjects at different frequencies, we explored whether signal was expressed at other crossmodulation frequencies (i.e. $n*40$ Hz $\pm$ $m*6$ Hz) in any individual subject based on visual inspection.

MRI processing, MEG-MRI coregistration and forward modeling

Structural MRI images were segmented and normalized to MNI space using unified segmentation (Ashburner and Friston, 2005). The inverse of this normalization transformation was employed to warp a template cortical mesh, i.e. a continuous tessellation of the cortex (excluding cerebellum) with 8196 vertices, from MNI space to each subject's native space. The MEG data were projected onto each subject's MRI space by applying a rigid body coregistration using the fiducials as landmarks. As head model, we employed a single shell aligned with the inner skull. Lead fields were then computed for each vertex in the cortical mesh with each dipole oriented normally to that mesh.

Model inversion

Source localization was performed within a parametric Bayesian framework using multiple sparse priors (MSP), with group-based optimized spatial priors (Mattout et al., 2007; Litvak and Friston, 2008). Before inversion, the data were projected to a subspace of 81 spatial modes based on a singular value decomposition (SVD) of the outer product of the leadfield matrix to retain >91.8% of the data variance. The projected data were further reduced to approximately 12.7 temporal modes (across subjects-mean) based on the SVD of the data variance matrix.

For each participant, MEG data for each condition and limited to different frequency bands (see below) were inverted separately using 1024 patches per hemisphere (plus 1024 bilateral patches) of the cortical mesh (smoothness parameter=0.6; estimated using greedy search algorithm). Additionally, anatomical priors were used that were defined by the AAL library (Tzourio-Mazoyer et al., 2002) using the MarsBaR toolbox (http://marsbar.sourceforge.net/) (Brett et al., 2002): Calcarine sulcus; Left Heschl's gyrus; Right Heschl's gyrus .

Source space analysis and statistics

For source space analysis, data were filtered to specific frequency bands of interest (see below) to optimize the computation of the inverse operator for those frequencies. Specifically, we applied the following filtering procedures: 1. To localize the sources of auditory and visual processing (including fundamental and higher order harmonics) and enable a direct comparison of auditory and visual SSRs, the data were broadband filtered to 1–90 Hz for reconstruction. 2. Non-specific audiovisual interplay (and the effect of synchrony) was characterized by comparing audiovisual with auditory (or visual) conditions at the specific fundamental frequencies of 6 Hz and 40 Hz. In other words, we investigated whether

(i) auditory processing at 40 Hz was amplified in the context of concurrent visual stimulation or (ii) visual processing at 6 Hz was amplified in the context of concurrent auditory stimulation. To enable a comparison between audiovisual and unisensory conditions selective for the fundamental frequencies that are present in both conditions (i.e. unconfounded by source activity generated in other frequency bands not present in both conditions), we narrowband filtered the data selectively around the sensory SSRs, i.e. 3–9 Hz for visual SSR and 37–43 Hz for auditory SSR.

Since no signal was detected at the across-sensory crossmodulation frequencies at the sensor level, no source localization was pursued for those frequency bands. Likewise, we did not pursue source identification of the within-sensory crossmodulation terms because of an insufficient signal to noise ratio. This is because the frequency resolution for the within-sensory crossmodulation terms (i.e. 0.1 Hz) can only be obtained by a trial length of greater than 10 s and thereby dramatically reduces the number of trials that can be used for averaging.

The filtered data were down sampled to 512 Hz and epoched into 1 second segments to obtain a higher signal to noise ratio (trading off against a lower frequency resolution than in sensor space analysis). After artefact rejection this resulted in an average of 470 epochs per condition and subject. The 1 second responses averaged across all epochs for each condition and subject were entered into the MSP inversion scheme to obtain source time-courses for each condition and subject at each point in the cortical mesh. The average energy of the source time course at the relevant frequencies were computed over the entire 1 second window; the source energies were interpolated into volumetric images in MNI space with 2 mm voxels and spatially smoothed with a 12 mm FWHM isotropic Gaussian kernel.

1) At the random effects or between-subject level, one source energy image (from the broadband filtered data) per condition and subject was entered into a repeated measures ANOVA modeling each of the 5 unisensory conditions in our experimental design and the subject-specific effects. In a conjunction null conjunction analysis, we compared visual and auditory unisensory conditions to identify auditory (and visual) evoked responses common to SM and DM conditions.

60

2) To evaluate non-specific effects of audiovisual interplay, source energy images for audiovisual and visual conditions (based on 3–9 Hz filtered data) were compared in paired t-tests separately for SM, DM1 or DM2 conditions. Likewise, source energy images for audiovisual and auditory conditions (based on 37–43 Hz filtered data) were compared in paired t-tests separately for SM, DM1 or DM2 conditions.

3) The effect of synchrony was investigated in paired t-tests comparing synchronous (i.e. DM1-AV) relative to asynchronous (i.e. DM2-AV) conditions (based on 3–9 Hz filtered data; n.b. 37–43 Hz filtered data are confounded because of different (i.e. 0.7 vs. 0.2 Hz) modulatory frequencies for DM1 and DM2 auditory stimulation.).

For a detailed overview of all statistical tests see Table 2.

Search volume constraints

The effects were tested for, in two nested search volumes (= anatomical regions of interest). The first search volume included the whole brain. The second search volume included either auditory (i.e. left and right Heschel's Gyrus (HG)) and/or visual (i.e. left and right calcarine sulcus, left and right cuneus, as well as the entire occipital cortex) cortices. Search volumes were defined by the AAL library (Tzourio-Mazoyer et al., 2002) using the MarsBaR toolbox (Brett et al., 2002) based on our a priori hypothesis that SSRs are located within unisensory cortices. Unless otherwise specified activations are reported at pb0.05 corrected at the peak level for multiple comparisons within the entire brain or the auditory and/or visual search volumes.

RESULTS

Sensor space results

Auditory and visual steady-state responses
Fig. 4 shows the frequency spectra for SM and DM2 audiovisual trials over all channels at a frequency resolution of 0.1 Hz for one representative subject. Peaks in amplitude, can be identified exactly at the auditory and visual stimulation frequencies and their higher harmonics. These effects were found consistently across all subjects. Indeed, the conjunction analysis identified a highly significant increase in amplitude for visual stimulation relative to fixation at 6 and 12 Hz and for auditory stimulation relative to fixation for 40 and 80 Hz (Table 1). The increases were widespread with a maximum over temporal sensors for auditory SSR frequencies and occipital sensors for visual SSR frequencies (Figs. 5A, 6A).

Signal integration within a sensory modality
As shown in Fig. 4, peaks in the frequency spectra were also found at within-sensory crossmodulation frequencies during double but not during single modulated trials. Specifically, we observed peaks at the sums and differences of (i) the frequency (40 Hz) and amplitude (0.7 Hz and 0.2 Hz) modulation of the double modulated auditory stimulus (e.g. DM2: 40 Hz±0.7 Hz=39.3 and 40.7 Hz) and (ii) the luminance (6 Hz) and size (0.2 Hz) modulation of the double modulated visual stimulus. Peaks at these crossmodulation frequencies were identified reliably in 13 of 14 subjects and were confirmed statistically. The peak amplitudes were increased at 5.8 and 6.2 Hz for DM-V relative to SM-V, at 39.8 and 40.2 Hz for DM1-A relative to SM-A and at 39.3 and 40.7 Hz for DM2-A relative to SM-A) (Table 1). Amplitude differences at within-sensory crossmodulation frequencies were located predominantly over temporal and occipital sensors (Figs. 5B, 6B).
In addition to crossmodulation frequencies as indices for true signal integration, we also observed an increase in amplitude at 40 Hz for DM2-A relative to SM-A trials (Table 1).

A. **Auditory & visual steady-state reponses**

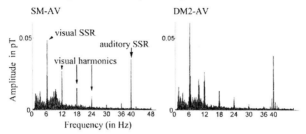

B. **Within-sensory frequency interactions**

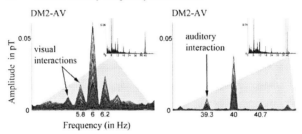

C. **Across-sensory frequency interactions**

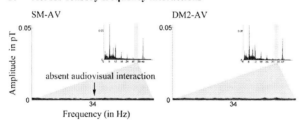

Figure 4. Frequency spectra for single and double modulated audiovisual trials. Mean magnitude (pT) as a function of frequency (Hz) is displayed for one representative subject (channel is colour coded). **(A)** Frequency spectrum for single (left) and double modulated (right) audiovisual conditions. **(B)** Enlarged frequency spectrum for double modulated audiovisual conditions (DM2-AV) cantered on within-sensory crossmodulation frequencies. Distinct peaks in amplitude are present at visual (e.g. 5.8 and 6.2 Hz) and auditory (e.g. 39.3 and 40.7 Hz) within crossmodulation frequencies. **(C)** Enlarged frequency spectrum for single modulated (left: SM-AV) and double modulated (right: DM2-AV) audiovisual trials cantered on the across-sensory crossmodulation frequency 34 Hz. No peaks in amplitude are visible

63

Table 1: Sensor space analysis: Statistical comparisons and results. Image coordinates (mm*mm*Hz) are given for the peak voxel. Sensor locations correspond to CTF conventions (first letter: R = right and L = left; second letter: T = temporal and O = occipital; numbering: 1st digit = row and 2nd digit = column). P-values are corrected for multiple comparisons for all voxels at frequencies of interest (FOIs).

Coordinates			Nearest sensor	Z-score (peak)	p-Value (peak)	Frequencies of interest (FOIs in Hz)
mm	mm	Hz				
Auditory & visual SSRs						
A > fixation (conjunction)						
57	-19	40.0	RT35	> 8	< 0.001	40, 80
47	-6	80.0	RT24	6.22	< 0.001	
V > fixation (conjunction)						
6	-73	6.0	RO21	> 8	< 0.001	6, 12
-4	-97	12.0	LO51	> 8	< 0.001	
Within-sensory signal integration						
DM-V ≠ SM-V						
4	-92	5.8	RO41	5.03	< 0.001	5.8, 6, 6.2, 12
4	-89	6.2	RO41	5.36	< 0.001	
DM1-A ≠ SM-A						
-43	-17	39.8	LT23	3.62	0.068	39.8, 40, 40.2, 80
-36	16	40.2	LF46	4.50	0.003	
DM2-A ≠ SM-A						
62	-36	39.3	RT53	3.97	0.023	
-4	-41	40.0	LP31	4.10	0.015	39.3, 40, 40.7, 80
30	5	40.7	RF65	4.40	0.005	
Across –sensory signal integration						
SM-AV ≠ SM-V						
-	-	-	-	< 3.1	n.s.	6,12, 28, 34, 46, 52
SM-AV ≠ SM-A						
-	-	-	-	< 3.1	n.s.	28, 34, 40, 46, 52, 80
DM1-AV ≠ DM-V						
-	-	-	-	< 3.1	n.s.	5.8, 6, 6.2, 12, 28, 34, 46, 52
DM2-AV ≠ DM-V						
-	-	-	-	< 3.1	n.s.	5.8, 6, 6.2, 12, 28, 34, 46, 52
DM1-AV ≠ DM1-A						
-	-	-	-	< 3.1	n.s.	28, 34, 39.8, 40, 40.2, 46, 52, 80
DM2-AV ≠ DM2-A						
-	-	-	-	< 3.1	n.s.	28, 34, 39.3, 40, 40.7, 46, 52, 80
Effect of synchronicity						
synchronous ≠ asynchrouns						
-	-	-	-	< 3.1	n.s.	5.8, 6, 6.2, 12, 40, 80

Signal integration across sensory modalities

Our data did not provide evidence for true signal integration across modalities as indicated by the absence of audiovisual crossmodulation frequencies in any single subject (Fig. 4C). Across subjects, no significant amplitude increases could be found at 28, 34, 46 or 52 Hz (i.e. 40 Hz±2*6 Hz) for audiovisual relative to unisensory trials. Even though we did not statistically investigate every possible crossmodulation frequency (i.e. n*40 Hz ± m*6 Hz), no evidence was provided by visual inspection of the data in any single subject either.

Moreover, comparing unisensory and bisensory trials did not identify any differences in amplitude at fundamental or harmonic frequencies as an index for possible lower level across-sensory saliency effects.

Finally, we did not observe any significant difference in amplitude at fundamental, harmonic or unisensory-visual crossmodulation frequencies between synchronous and asynchronous trials.

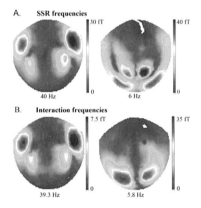

Figure. 5. (A) Topographies of signal mean magnitude at the auditory (40 Hz, left) and visual (6 Hz, right) stimulation frequencies, in one representative subject. **(B)**, Topographies of signal mean magnitude at the auditory (39.3 Hz, left) and visual (5.8 Hz, right) within-sensory crossmodulation frequencies.

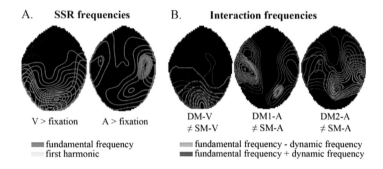

A. SSR frequencies   B. Interaction frequencies

V > fixation   A > fixation   DM-V   DM1-A   DM2-A
                                 ≠ SM-V   ≠ SM-A   ≠ SM-A

■ fundamental frequency   ■ fundamental frequency - dynamic frequency
   first harmonic   ■ fundamental frequency + dynamic frequency

Figure 6. Topographies of the sensor level statistics. **(A)** Visual SSRs>Fixation (left) and Auditory SSRs>Fixation at fundamental frequencies, i.e. 6 and 40 Hz, (blue) and first harmonics, i.e. 12 and 80 Hz (yellow). Height threshold: t>5.9, pFWEb0.05. **(B)** Within-sensory crossmodulation frequencies. Left: DM-V≠SM-V at 5.8 (green) and 6.2 Hz (red); Middle: DM1- A≠SM-A at 39.8 (green) and 40.2 Hz (red); Right: DM2-A≠SM-A at 39.3 (green) and 40.7 Hz (red). Height threshold: F>18, pb0.001 (uncorrected). (For interpretation of the references to colour in this figure legend, the reader is referred to the web version of this article.)

Source space results

Auditory and visual steady-state responses

Visual relative to auditory SSR stimulation induced significant increases in source energy within bilateral occipital cortices. The maximal source energy for SSRs was found close to the primary visual cortex in descendens gyrus/middle occipital gyrus extending in the cuneus, calcarine gyrus, inferior/middle/superior occipital gyri and the lingual gyri of both hemispheres (Table 2, Fig. 7B).

Auditory relative to visual stimulation increases in source energy could be found within left and right rolandic opercula (i.e. subcentral gyri) extending into left and right postcentral gyri and left temporal gyrus. Specifically, the peaks in source energy were found in the right and left rolandic operculum and extended into HG in the left hemisphere (Table 2, Fig. 7A).

Multisensory vs. unisensory SSRs

Since no signal could be identified at any audiovisual crossmodulation frequencies in sensor space, we investigated effects of audiovisual stimulation only by comparing mean evoked source energy at the harmonic frequencies of auditory or visual stimulation (i.e. only non-specific audiovisual interplay). Indeed for the data filtered around 6 Hz (3–9 Hz), the mean evoked source energy was enhanced in the left middle/inferior occipital gyrus and right middle occipital gyrus for DM1-AV (i.e. synchronous) trials relative to DM-V trials (Table 2, Fig. 7C). For the data filtered around 40 Hz (37–43 Hz), however, no significant effects were found.

Effects of synchrony

Comparing the mean evoked source energy at 6 Hz for synchronous and asynchronous audiovisual stimulation demonstrated no significant differences.

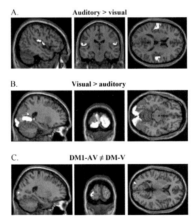

Figure 7. Source space results displayed on a canonical MNI brain. Increases in mean source energy across subjects for auditory relative to visual **(A)** and visual relative to auditory **(B)** conditions. Height threshold: t>3.25, pb0.001 uncorrected. Extent threshold: >100 voxels. **(C)** Differences in mean source energy across subjects for synchronous audiovisual vs. unisensory visual conditions. Height threshold: F>17.82, pb0.001 uncorrected. Extent threshold: >50 voxels.

Table 2: Source space analysis. All statistical comparisons and results. P-values are corrected at the peak level for multiple comparisons within the entire brain or the visual (*) or auditory (†) search volumes.

Brain Region	MNI Coordinates			Cluster size	Z-score (peak)	p-Value (peak)
	x	y	z			
Auditory & visual source energy						
A > V (conjunction)						
L. HG	- 48	- 10	10	4	4.00	0.003†
L. subcentral gyrus (rolandic operculum)	- 46	- 18	16	346	5.08	0.005
R. subcentral gyrus (rolandic operculum)	50	-10	14	220	4.12	0.228
V > A (conjunction)						
L. gyrus descendens (Ecker) /middle occipital gyrus	- 22	- 96	- 12	24562	> 8	< 0.001
Across-sensory source energy						
SM-AV ≠ SM-V (3-9 Hz filter)						
-	-	-	-	-	-	n.s.
DM1-AV ≠ DM-V (3-9 Hz filter)						
R. middle occipital gyrus	- 22	- 96	6	184	4.39	0.014*
L. middle/inferior occipital gyrus	26	- 90	0	77	4.19	0.028*
DM2-AV ≠ DM-V (3-9 Hz filter)						
-	-	-	-	-	-	n.s.
SM-AV ≠ SM-A (40 Hz filter)						
-	-	-	-	-	-	n.s.
DM1-AV ≠ DM-A (37-43 Hz filter)						
-	-	-	-	-	-	n.s.
DM2-AV ≠ DM-A (37-43 Hz filter)						
-	-	-	-	-	-	n.s.
Effect of synchronicity						
DM1-AV ≠ DM2-AV (37-43 Hz filter)						
-	-	-	-	-	-	n.s.
-	-	-	-	-	-	n.s.

† Small volume corrected for the auditory search volume
* Small volume corrected for the visual search volume

DISCUSSION

This MEG study investigated signal integration within and across the auditory and visual modalities by analysing steady-state responses in the frequency domain. Our results suggest that stimulus features at different temporal scales are integrated within but not across sensory modalities.

Steady-state responses

Previous research suggested that tones modulated in amplitude or frequency at 40 Hz elicit reliable auditory steady-state responses (Galambos et al., 1981; Picton et al., 1987) Using dipole modeling and beamformer methods, the auditory SSRs were localized to Heschl's gyri (Schoonhoven et al., 2003; Millman et al., 2010; Okamoto et al., 2010; Steinmann and Gutschalk, 2011). In comparison with the N1 response, dipole localization even suggested that the auditory SSR generators are located more medially in primary auditory cortex (Ross et al., 2003; Okamoto et al., 2010). A location of auditory SSR in primary auditory cortex was further supported by invasive human electrophysiology (Bidet-Caulet et al., 2007; Brugge et al., 2009). In line with previous research, our 40 Hz frequency modulated tone also elicited a robust auditory SSR that could be identified reliably not only at the group level, but also in every single subject at the sensor level. Subsequent distributed source analysis based on multiple sparse priors located the neural activity medially in the Heschl's gyri or the parietal operculum covering Heschl's gyrus.

Likewise, manipulating the grating's luminance at 6 Hz elicited a visual SSR reliably at the group level and in every single subject. Distributed source analysis demonstrated widespread activation within the calcarine sulci extending into the occipito-temporal cortices. These source findings converge with previous dipole localization results suggesting that the VSSR is generated by multiple sources including the primary visual cortices and areas V5/hMT+ bilaterally (Pastor et al., 2003; Fawcett et al., 2004; Di Russo et al., 2007).

69

Interactions within sensory modalities

In the double modulated conditions, auditory and visual stimuli were modulated concurrently at two different frequencies. This enabled us to investigate how stimulus features are integrated within a single sensory modality across different temporal scales.

In the auditory modality, environmental sounds and speech are endowed with a complex temporal structure. In particular, speech carries complementary and partly redundant information at several temporal scales (Rosen, 1992; Shannon et al., 1995). While speech intelligibility is well preserved when the stimulus is reduced to slow amplitude modulations, additional fast modulations of the so-called fine structure render speech much more robust against corruptions from background (Zeng et al., 2005). Previous research has focused primarily either on the effect of imposing multiple amplitude modulations to tones with the identical or different carrier frequencies (Lins and Picton, 1995; John et al., 1998; Draganova et al., 2002; Ross et al., 2003; Draganova et al., 2008) or on the effect of imposing slow frequency modulations on fast amplitude modulations (Luo et al., 2006, 2007). In contrast, we investigated the effect of slow (i.e. 0.7 or 0.3 Hz) amplitude modulation on 40 Hz frequency modulations. In line with one recent study (Ding and Simon, 2009), simultaneous modulation of a tone's amplitude and frequency induced crossmodulation frequencies at the sums and differences of the stimulation frequencies indicating non-linear integration of frequency and amplitude information across different time scales.

In the visual modality, previous steady-state stimulation paradigms have used crossmodulation frequencies to demonstrate interactions between multiple stimuli at overlapping or disparate regions in space (Ratliff and Zemon, 1982; Regan and Regan, 1988b; Fuchs et al., 2008; Sutoyo and Srinivasan, 2009). We investigated whether stimulus features such as a grating's luminance and size are integrated across different time scales. Indeed, simultaneous modulations of a grating's luminance and size at different frequencies induced pronounced nonlinear crossmodulation frequencies in support of feature integration and binding. Importantly, at least one crossmodulation frequency could be identified reliably in every single subject.

Interactions across sensory modalities

Over the past decade accumulating evidence has challenged the traditional late integration model. Collectively, evidence from neuroanatomy, neurophysiology and functional imaging in humans suggests that multisensory integration starts early at about 50–100 ms post stimulus and emerges already at the primary cortical level (Molholm et al., 2002; Schroeder and Foxe, 2002; Kayser et al., 2005; Talsma et al., 2007; Werner and Noppeney, 2010a). This raises the question whether multisensory integration can be identified in terms of crossmodulation frequencies when the human brain is stimulated with steady-state stimuli in the auditory and visual modalities. To our knowledge, only one previous study has specifically investigated and reported crossmodulation frequencies for auditory and visual stimuli modulated at different frequencies (Regan et al., 1995). Since the methods sections of this previous research article is relatively sparse and does not clearly define the number of subjects or statistical procedures, the current study aimed to replicate their findings. Yet, in contrast to the robust within-modality crossmodulation frequencies that could be identified reliably at the single subject level, none of the subjects showed audiovisual crossmodulation frequencies in the single or double modulated trials. Further, crossmodulation frequencies for cross-sensory integration were not observed in any of our exploratory studies that manipulated the modulation frequencies, the length of the trials or subjects' attentional set. Even when both auditory and visual signals were additionally slowly modulated at 0.2 Hz and thereby created the percept of synchronous looming and receding, no audiovisual interactions were observed.

Obviously, null-results always need to be interpreted with great caution. Nevertheless, the contrast between the robust identification of within-modality crossmodulation frequencies and the equally reliable absence of across-modality effects in every single subject suggests that information modulated in a steady fashion at different frequencies is integrated within but not across the senses. The absence of audiovisual integration responses contrasts with the numerous previous EEG studies demonstrating audiovisual interactions in ERP responses (Teder-Salejarvi et al., 2002; Talsma and Woldorff, 2005; Teder-Salejarvi et al., 2005; Talsma et al., 2007; Raij et al., 2010). This pattern of results suggests that audiovisual interactions at the primary cortical level may crucially depend on rapid stimulus transients that co-occur in space and time in auditory and visual senses. In line with this conjecture, a recent fMRI study dissociating transient and sustained BOLD responses demonstrated that primary auditory and visual areas integrated only the rapid stimulus transients at stimulus onset (Werner and Noppeney, 2010a). In contrast, audiovisual integration for sustained responses and slow modulations (0.1 Hz) emerged in higher order association areas. These findings suggest that the temporal window for multisensory integration and the type of integration may differ along the cortical hierarchy. Primary sensory cortices may have narrow temporal integration windows ideal for temporal coincidence detection and integration of brief transients thereby enabling an initial scene segmentation. Indeed, it is well established that integration at the primary cortical level is sensitive to even small degrees of temporal asynchrony (Schroeder and Foxe, 2002; Kayser et al., 2005; Ghazanfar and Schroeder, 2006; Lewis and Noppeney, 2010). By contrast, our steady state paradigm provides precise temporal coincidence cues only at the on- and offsets of each 60 second stimulation period. Hence, steady state responses that emerge primarily in primary sensory cortices may not be suitable to reveal the audiovisual integration processes critical for low level coincidence detection.

While we did not observe crossmodulation frequencies as an index for true audiovisual integration, source level analysis identified differences in source energy in the occipital cortices for audiovisual DM synchronous relative to the corresponding unisensory synchronous conditions. These effects may have been attenuated at the sensor level, because of superposition of activity from multiple sources, intersession and inter-subject variability that our analysis cannot account for at the sensor level. Non-specific differences between audiovisual and unisensory conditions indicate that synchronous audiovisual stimulation enhances the bottom-up saliency of the incoming sensory inputs or modulates subjects' cognitive set. For instance, concurrent presentation of synchronous visual and auditory looming stimuli in our study may have increased subjects' attentional level.

Conclusions

Over the past decade, evidence has accumulated for audiovisual interactions starting already at the primary cortical level. Using auditory and visual SSRs, we did not identify non-linear audiovisual interactions as indexed by crossmodulation frequencies at the group or single subject level. The absence of audiovisual crossmodulation frequencies suggests that the previously reported audiovisual interactions in primary sensory areas may mediate low level spatiotemporal coincidence detection that is prominent for stimulus transients but less relevant for sustained SSR responses. In contrast, crossmodulation frequencies were observed when two stimulus features within the same sensory modality were modulated at different frequencies suggesting integration of stimulus features within one sensory modality across different time scales already at the primary cortical level.

In conclusion, our results demonstrate that information represented in SSR is integrated at the primary cortical level within but not across sensory modalities.

REFERENCES

Ashburner J, Friston KJ (2005) Unified segmentation. Neuroimage 26:839-851.

Beauchamp MS, Argall BD, Bodurka J, Duyn JH, Martin A (2004) Unraveling multisensory integration: patchy organization within human STS multisensory cortex. Nat Neurosci 7:1190-1192.

Bidet-Caulet A, Fischer C, Besle J, Aguera PE, Giard MH, Bertrand O (2007) Effects of selective attention on the electrophysiological representation of concurrent sounds in the human auditory cortex. J Neurosci 27:9252-9261.

Brainard DH (1997) The Psychophysics Toolbox. Spat Vis 10:433-436.

Brett M, Anton J, Valabregue R, Poline JB (2002) Region of interest analysis using an SPM toolbox. Neuroimage 16:Presented at the 8th International Conference on Functional Mapping of the Human Brain, June 2-6, 2002, Sendai, Japan.

Brugge JF, Nourski KV, Oya H, Reale RA, Kawasaki H, Steinschneider M, Howard MA, 3rd (2009) Coding of repetitive transients by auditory cortex on Heschl's gyrus. J Neurophysiol 102:2358-2374.

Calvert GA (2001) Crossmodal processing in the human brain: insights from functional neuroimaging studies. Cereb Cortex 11:1110-1123.

Cosmelli D, David O, Lachaux JP, Martinerie J, Garnero L, Renault B, Varela F (2004) Waves of consciousness: ongoing cortical patterns during binocular rivalry. Neuroimage 23:128-140.

de Jong R, Toffanin P, Harbers M (2010) Dynamic crossmodal links revealed by steady-state responses in auditory-visual divided attention. Int J Psychophysiol 75:3-15.

Di Russo F, Pitzalis S, Aprile T, Spitoni G, Patria F, Stella A, Spinelli D, Hillyard SA (2007) Spatiotemporal analysis of the cortical sources of the steady-state visual evoked potential. Hum Brain Mapp 28:323-334.

Dimitrijevic A, John MS, van Roon P, Picton TW (2001) Human auditory steady-state responses to tones independently modulated in both frequency and amplitude. Ear Hear 22:100-111.

Ding N, Simon JZ (2009) Neural representations of complex temporal modulations in the human auditory cortex. J Neurophysiol 102:2731-2743.

Draganova R, Ross B, Borgmann C, Pantev C (2002) Auditory cortical response patterns to multiple rhythms of AM sound. Ear Hear 23:254-265.

Draganova R, Ross B, Wollbrink A, Pantev C (2008) Cortical steady-state responses to central and peripheral auditory beats. Cereb Cortex 18:1193-1200.

Duncan J, Martens S, Ward R (1997) Restricted attentional capacity within but not between sensory modalities. Nature 387:808-810.

Fawcett IP, Barnes GR, Hillebrand A, Singh KD (2004) The temporal frequency tuning of human visual cortex investigated using synthetic aperture magnetometry. Neuroimage 21:1542-1553.

Fort A, Delpuech C, Pernier J, Giard MH (2002) Dynamics of cortico-subcortical cross-modal operations involved in audio-visual object detection in humans. Cereb Cortex 12:1031-1039.

Friston KJ, Penny WD, Glaser DE (2005) Conjunction revisited. Neuroimage 25:661-667.

Fuchs S, Andersen SK, Gruber T, Muller MM (2008) Attentional bias of competitive interactions in neuronal networks of early visual processing in the human brain. Neuroimage 41:1086-1101.

Galambos R, Makeig S, Talmachoff PJ (1981) A 40-Hz auditory potential recorded from the human scalp. Proc Natl Acad Sci U S A 78:2643-2647.

Gander PE, Bosnyak DJ, Roberts LE (2010) Evidence for modality-specific but not frequency-specific modulation of human primary auditory cortex by attention. Hear Res 268:213-226.

Ghazanfar AA, Schroeder CE (2006) Is neocortex essentially multisensory? Trends Cogn Sci 10:278-285.

Ghazanfar AA, Chandrasekaran C, Logothetis NK (2008) Interactions between the superior temporal sulcus and auditory cortex mediate dynamic face/voice integration in rhesus monkeys. J Neurosci 28:4457-4469.

Giard MH, Peronnet F (1999) Auditory-visual integration during multimodal object recognition in humans: a behavioral and electrophysiological study. J Cogn Neurosci 11:473-490.

Goebel R, van Atteveldt N (2009) Multisensory functional magnetic resonance imaging: a future perspective. Exp Brain Res 198:153-164.

Hari R, Hamalainen M, Joutsiniemi SL (1989) Neuromagnetic steady-state responses to auditory stimuli. J Acoust Soc Am 86:1033-1039.

Hillis JM, Ernst MO, Banks MS, Landy MS (2002) Combining sensory information: mandatory fusion within, but not between, senses. Science 298:1627-1630.

John MS, Lins OG, Boucher BL, Picton TW (1998) Multiple auditory steady-state responses (MASTER): stimulus and recording parameters. Audiology 37:59-82.

Kamphuisen A, Bauer M, van Ee R (2008) No evidence for widespread synchronized networks in binocular rivalry: MEG frequency tagging entrains primarily early visual cortex. J Vis 8:4 1-8.

Kayser C, Petkov CI, Augath M, Logothetis NK (2005) Integration of touch and sound in auditory cortex. Neuron 48:373-384.

Kayser C, Petkov CI, Augath M, Logothetis NK (2007) Functional imaging reveals visual modulation of specific fields in auditory cortex. J Neurosci 27:1824-1835.

Kleiner M, Brainard D, Pelli D (2007) What's new in Psychtoolbox-3? Perception 36.

Lakatos P, Chen CM, O'Connell MN, Mills A, Schroeder CE (2007) Neuronal oscillations and multisensory interaction in primary auditory cortex. Neuron 53:279-292.

Lewis R, Noppeney U (2010) Audiovisual synchrony improves motion discrimination via enhanced connectivity between early visual and auditory areas. J Neurosci 30:12329-12339.

Lins OG, Picton TW (1995) Auditory steady-state responses to multiple simultaneous stimuli. Electroencephalogr Clin Neurophysiol 96:420-432.

Litvak V, Friston K (2008) Electromagnetic source reconstruction for group studies. Neuroimage 42:1490-1498.

Luo H, Wang Y, Poeppel D, Simon JZ (2006) Concurrent encoding of frequency and amplitude modulation in human auditory cortex: MEG evidence. J Neurophysiol 96:2712-2723.

Luo H, Wang Y, Poeppel D, Simon JZ (2007) Concurrent encoding of frequency and amplitude modulation in human auditory cortex: encoding transition. J Neurophysiol 98:3473-3485.

Mattout J, Henson RN, Friston KJ (2007) Canonical source reconstruction for MEG. Comput Intell Neurosci:67613.

Millman RE, Prendergast G, Kitterick PT, Woods WP, Green GG (2010) Spatiotemporal reconstruction of the auditory steady-state response to frequency modulation using magnetoencephalography. Neuroimage 49:745-758.

Molholm S, Ritter W, Murray MM, Javitt DC, Schroeder CE, Foxe JJ (2002) Multisensory auditory-visual interactions during early sensory processing in humans: a high-density electrical mapping study. Brain Res Cogn Brain Res 14:115-128.

Nichols T, Brett M, Andersson J, Wager T, Poline JB (2005) Valid conjunction inference with the minimum statistic. Neuroimage 25:653-660.

Noesselt T, Bonath B, Boehler CN, Schoenfeld MA, Heinze HJ (2008) On perceived synchrony-neural dynamics of audiovisual illusions and suppressions. Brain Res 1220:132-141.

Noppeney U (2011) Characterization of multisensory integration with fMRI: Experimental design, statistical analysis and interpretation. In: Frontiers in the Neural Basis of Multisensory Processes. Boca Raton, FL, USA: CRC Press.

Noppeney U, Ostwald D, Werner S (2010) Perceptual decisions formed by accumulation of audiovisual evidence in prefrontal cortex. J Neurosci 30:7434-7446.

Okamoto H, Stracke H, Bermudez P, Pantev C (2010) Sound Processing Hierarchy within Human Auditory Cortex. J Cogn Neurosci.

Pastor MA, Artieda J, Arbizu J, Valencia M, Masdeu JC (2003) Human cerebral activation during steady-state visual-evoked responses. J Neurosci 23:11621-11627.

Picton TW, John MS, Dimitrijevic A, Purcell D (2003) Human auditory steady-state responses. Int J Audiol 42:177-219.

Picton TW, Skinner CR, Champagne SC, Kellett AJ, Maiste AC (1987) Potentials evoked by the sinusoidal modulation of the amplitude or frequency of a tone. J Acoust Soc Am 82:165-178.

Raij T, Ahveninen J, Lin FH, Witzel T, Jaaskelainen IP, Letham B, Israeli E, Sahyoun C, Vasios C, Stufflebeam S, Hamalainen M, Belliveau JW (2010) Onset timing of cross-sensory activations and multisensory interactions in auditory and visual sensory cortices. Eur J Neurosci 31:1772-1782.

Ratliff F, Zemon V (1982) Some new methods for the analysis of lateral interactions that influence the visual evoked potential. Ann N Y Acad Sci 388:113-124.

Regan D, Spekreijse H (1977) Auditory-visual interactions and the correspondence between perceived auditory space and perceived visual space. Perception 6:133-138.

Regan D, Regan MP (1988a) The transducer characteristic of hair cells in the human ear: a possible objective measure. Brain Res 438:363-365.

Regan D, Regan MP (1988b) Objective evidence for phase-independent spatial frequency analysis in the human visual pathway. Vision Res 28:187-191.

Regan MP, He P, Regan D (1995) An audio-visual convergence area in the human brain. Exp Brain Res 106:485-487.

Rosen S (1992) Temporal Information in Speech - Acoustic, Auditory and Linguistic Aspects. Philos Trans R Soc Lond B Biol Sci 336:367-373.

Ross B, Draganova R, Picton TW, Pantev C (2003) Frequency specificity of 40-Hz auditory steady-state responses. Hear Res 186:57-68.

Ross B, Borgmann C, Draganova R, Roberts LE, Pantev C (2000) A high-precision magnetoencephalographic study of human auditory steady-state responses to amplitude-modulated tones. J Acoust Soc Am 108:679-691.

Sadaghiani S, Maier JX, Noppeney U (2009) Natural, metaphoric, and linguistic auditory direction signals have distinct influences on visual motion processing. J Neurosci 29:6490-6499.

Sarkheil P, Vuong QC, Bulthoff HH, Noppeney U (2008) The integration of higher order form and motion by the human brain. Neuroimage 42:1529-1536.

Saupe K, Schroger E, Andersen SK, Muller MM (2009a) Neural mechanisms of intermodal sustained selective attention with concurrently presented auditory and visual stimuli. Front Hum Neurosci 3:58.

Saupe K, Widmann A, Bendixen A, Muller MM, Schroger E (2009b) Effects of intermodal attention on the auditory steady-state response and the event-related potential. Psychophysiology 46:321-327.

Schoonhoven R, Boden CJ, Verbunt JP, de Munck JC (2003) A whole head MEG study of the amplitude-modulation-following response: phase coherence, group delay and dipole source analysis. Clin Neurophysiol 114:2096-2106.

Schroeder CE, Foxe JJ (2002) The timing and laminar profile of converging inputs to multisensory areas of the macaque neocortex. Brain Res Cogn Brain Res 14:187-198.

Schroger E, Widmann A (1998) Speeded responses to audiovisual signal changes result from bimodal integration. Psychophysiology 35:755-759.

Seymour K, Clifford CW, Logothetis NK, Bartels A (2010) Coding and binding of color and form in visual cortex. Cereb Cortex 20:1946-1954.

Shannon RV, Zeng FG, Kamath V, Wygonski J, Ekelid M (1995) Speech recognition with primarily temporal cues. Science 270:303-304.

Srinivasan R (2004) Internal and external neural synchronization during conscious perception. Int J Bifurcat Chaos 14:825-842.

Steinmann I, Gutschalk A (2011) Potential fMRI correlates of 40-Hz phase locking in primary auditory cortex, thalamus and midbrain. Neuroimage 54:495-504.

Sutoyo D, Srinivasan R (2009) Nonlinear SSVEP responses are sensitive to the perceptual binding of visual hemifields during conventional 'eye' rivalry and interocular 'percept' rivalry. Brain Res 1251:245-255.

Talsma D, Woldorff MG (2005) Selective attention and multisensory integration: multiple phases of effects on the evoked brain activity. J Cogn Neurosci 17:1098-1114.

Talsma D, Doty TJ, Woldorff MG (2007) Selective attention and audiovisual integration: is attending to both modalities a prerequisite for early integration? Cereb Cortex 17:679-690.

Teder-Salejarvi WA, McDonald JJ, Di Russo F, Hillyard SA (2002) An analysis of audio-visual crossmodal integration by means of event-related potential (ERP) recordings. Brain Res Cogn Brain Res 14:106-114.

Teder-Salejarvi WA, Di Russo F, McDonald JJ, Hillyard SA (2005) Effects of spatial congruity on audio-visual multimodal integration. J Cogn Neurosci 17:1396-1409.

Tononi G, Srinivasan R, Russell DP, Edelman GM (1998) Investigating neural correlates of conscious perception by frequency-tagged neuromagnetic responses. Proc Natl Acad Sci U S A 95:3198-3203.

Tzourio-Mazoyer N, Landeau B, Papathanassiou D, Crivello F, Etard O, Delcroix N, Mazoyer B, Joliot M (2002) Automated anatomical labeling of activations in SPM using a macroscopic anatomical parcellation of the MNI MRI single-subject brain. Neuroimage 15:273-289.

van Atteveldt N, Formisano E, Goebel R, Blomert L (2004) Integration of letters and speech sounds in the human brain. Neuron 43:271-282.

Vialatte FB, Maurice M, Dauwels J, Cichocki A (2010) Steady-state visually evoked potentials: focus on essential paradigms and future perspectives. Prog Neurobiol 90:418-438.

Wallace MT, Wilkinson LK, Stein BE (1996) Representation and integration of multiple sensory inputs in primate superior colliculus. J Neurophysiol 76:1246-1266.

Wallace MT, Roberson GE, Hairston WD, Stein BE, Vaughan JW, Schirillo JA (2004) Unifying multisensory signals across time and space. Exp Brain Res 158:252-258.

Werner S, Noppeney U (2010a) Distinct functional contributions of primary sensory and association areas to audiovisual integration in object categorization. J Neurosci 30:2662-2675.

Werner S, Noppeney U (2010b) The Contributions of Transient and Sustained Response Codes to Audiovisual Integration. Cereb Cortex.

Zeng FG, Nie K, Stickney GS, Kong YY, Vongphoe M, Bhargave A, Wei C, Cao K (2005) Speech recognition with amplitude and frequency modulations. Proc Natl Acad Sci U S A 102:2293-2298.

CHAPTER 3:

FROM COMPLEX AUDITORY SCENES TO PERCEPTUAL AWARENESS WITHIN 600 MS

Anette S. Giani[1], Paolo Belardinelli[2], Erick Ortiz[2], Mario Kleiner[1] & Uta Noppeney[1,3]

[1] Max Planck Institute for Biological Cybernetics, Tübingen, Germany

[2] MEG Centre, University of Tübingen, Tübingen, Germany

[3] Computational Neuroscience and Cognitive Robotics Centre, University of Birmingham, UK

ABSTRACT

In daily life, our auditory system detects and segregates sounds, derived from complex auditory scenes. Yet, limited processing capacities allow only a small subset of these sounds to enter awareness. This magnetoencephalography (MEG) study used informational masking to identify the neural mechanisms that enable auditory awareness. On each trial, participants indicated whether they detected a sequence of two target tones that were embedded within a multi-tone background. Hence, target detection required participants to sequentially integrate both target tones in time. We analysed MEG activity for 'hits' and 'misses', separately for the first and second target in a sequence. Comparing stimuli that were physically identical, yet perceptually different, allowed us to assess neural processing underlying participants' awareness. Results showed that awareness of a two-tone sequence entails complex neural dynamics that depend on the tone's sequential position. In particular, awareness selectively enhanced early processing of the first target, probably reflecting sensory gating mechanisms. Moreover, it correlated with an awareness related negativity (ARN) that occurred after the second target tone. This ARN thus followed the sequential integration of both target tones and relied on intrinsic modulation within the auditory cortex. Lastly, we found a sustained deflection peaking around 300 and 500 ms after the onset of both tones, probably reflecting working memory processing that typically precede awareness. We showed that the sustained activity depended on recurrent modulations of the connections between auditory and parietal cortices. Concluding, our results suggest that awareness of a two-tone pair relies on a cascade of processes within a temporo-parietal network.

INTRODUCTION

In our daily life, we are constantly bombarded with many environmental sounds which sum up to a single complex sound wave. With seemingly no effort, our auditory system breaks down this sound wave and integrates information into coherent streams. Yet, only a fraction of these streams enter our awareness. How does the human brain enable perceptual awareness within complex auditory scenes?

Perceptual awareness is believed to rely on activations within a global workspace network (reviewed in: Dehaene et al., 2006). This network crucially depends on recurrent activation in early sensory cortices (Lamme and Roelfsema, 2000; Ro et al., 2003; Gutschalk et al., 2008) and top-down amplification of parieto-frontal areas (Dehaene et al., 2006). However, these theoretical frameworks are based predominantly on studies that investigated visual awareness.

Crucially, auditory awareness relies on stream segregation: An auditory signal that evolves over time is segmented from background (Bregman, 1990). To study this process, informational masking has proven a powerful paradigm. In informational masking experiments, participants typically have to detect a sequence of pure-tone targets embedded within a multi-tone mask (Neff and Green, 1987). Stimulus detection thus requires participants to identify the sequence of target tones via temporal integration and stream segregation. Critically, target tones are presented in a protected region in frequency space thus avoiding overlap of sound energies of masking and target tones (Durlach et al., 2003).

Based on informational masking, previous MEG studies suggested that auditory awareness is associated with an awareness related negativity (ARN). The ARN emerged in auditory cortex about 150 ms after stimulus onset (Gutschalk et al., 2008; Konigs and Gutschalk, 2012; Wiegand and Gutschalk, 2012). By contrast, early (< 100 ms) evoked components and steady-state responses (SSRs) were not influenced by awareness. This dissociation led the authors to conclude that auditory awareness emerges between early and late processing stages in auditory cortex (Gutschalk et al., 2008).

Critically however, authors based this conclusion on evoked activity that was pooled over multiple tones within a sequence. However, since a target was defined as a sequential stream

81

of two (or more) tones. Indeed, the build-up of auditory streams is accompanied by changes in neural processing (Cusack et al., 2004).

Previous informational masking studies, too, did not identify any later (> 250 ms) awareness effects. This is remarkable, because the later, recurrent modulations within a fronto-parietal network appear crucial for perceptual awareness (Dehaene et al., 2006) and auditory stream formation (Cusack, 2005; Teki et al., 2011).

This MEG study was designed to gain further insights into the *complete* neuronal cascade, which enables auditory awareness. Using informational masking we dissociated neural activity associated with processing the first and second tone. We further investigated the influence of recurrent activity within the auditory-parietal network using dynamic causal modelling (DCM).

MATERIALS AND METHODS

Participants

After giving informed consent, 21 healthy young adults participated in this study (12 females, 20 right-handed, mean age (± standard deviation): 26.2 ± 4.04 years, range: 20-36 years). All of them reported normal hearing and had normal or corrected-to-normal vision. Data from one female participant was discarded from the analysis due to excessive eye blinks. The study was approved by the local ethics review board of the University of Tübingen.

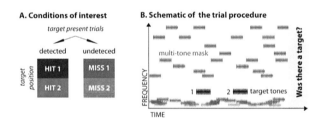

Figure 1. Experimental design and conditions. (**A**) Conditions of interest. We focused only on target present trials that were sorted post-hoc into 'hits' and 'misses'. Each target tone within a sequence was analysed separately. Hence, our experimental paradigm conformed to a 2 x 2 factorial design with the factors: (1) awareness (hits, misses) and (2) target position (first, second). (**B**) The informational masking paradigm. Participants detected a pair of target tones (black) embedded within a multi-tone mask (grey). Target tones were presented in a protected region in frequency space (grey shading). After each trial participants indicated whether there was a target pair.

Experimental Design

On each trial, participants detected a pair of target tones that was embedded within a multi-tone mask (**Figure 1B**). The target pair was presented with a fixed stimulus onset asynchrony. It was present in 2/3 of the trials (target present trials) and absent in the remaining 1/3 (target absent or catch trials). In additional trials, visual gratings were presented together with the auditory targets; however, these trials are not included or discussed in this report.

In contrast, this report focusses only on the auditory target present trials that were categorized post hoc into 'hits' and 'misses' according to participants' responses. Comparing hits and misses to identical auditory stimuli enabled us to assess the role of participants' awareness on auditory processing. Furthermore, we evaluated the effect of awareness for the first and second target tone separately to characterize the cascade of processes that enable a pair of tones to be segmented from a complex auditory scene and enter our awareness. Hence, our experimental paradigm conformed to a 2 x 2 factorial design with the factors: (1) *awareness* (hit, miss) and (2) *target-position* (first, second) (**Figure 1A**). Please note that even though we will analyse and report the MEG responses to the 1st and 2nd target tones separately, participants needed to detect the two target tones as a pair. Hence, on a particular trial, both targets were always classified together as hits or misses.

Schematic of a trial procedure

At the beginning of each trial there was a random delay period of 800 – 1300 ms (mean 1050 ms). Subsequently, on target present trials, the first target tone was played for a period of 300 ms. After a fixed stimulus onset asynchrony (SOA) of 1050 ms the second auditory target was presented. 1050 ms after onset of the second target the question: "Was there an auditory target?" appeared on the screen. Participants indicated their response via a two choice key press with their right index or middle finger (order randomized) within a maximal response time of 2 seconds. A visual fixation cross and a multi-tone mask were present throughout the entire trial.

Stimuli

All target and masking tones were amplitude modulated tones with a duration of 300 ms and a carrier frequency selected from a set of 26 frequencies. These frequencies were equally spaced on a logarithmic scale ranging from 200 to 5065 Hz.

On each trial, one single carrier frequency was selected commonly for both of the target tones from a set of five target frequencies: 1222, 1583, 2049, 2654 and 3437 Hz. This target frequency and additional 3 frequencies above and below the target frequency were then protected from being used as masking tones. Hence, the multi-tone, informational mask included only tones of the remaining 19 frequencies. To maximize variability of the masking tones, these frequencies varied around one estimated rectangular bandwidth [ERB = 24.7 x (4.37 x fc + 1)] (Moore, 1995; Gutschalk et al., 2008). The SOA of the masking tones was randomized within each frequency band between 550 and 1550ms (mean 1050 ms), excluding a protected region of 850 to 1250 ms centred around the fixed target tones' SOA of 1050 ms.

To evoke SSRs the target tones' amplitude was modulated sinusoidally at a rate of 40 Hz and a modulation depth of 100% which has consistently been shown to evoke robust auditory SSRs (for a review see: Picton et al., 2003). The masking tones' amplitude was modulated sinusoidally at rates of 32, 36, 44 or 48 Hz, thus enhancing the similarity between masking and target tones. To avoid clicking sounds, the masking tones were multiplied with 10 ms sinusoidal ramps at on- and offsets. Target tones were presented at a mean loudness of 50.5 dB sensation level (SL), while masking tones level was, on average, 4 dB SL louder.

All auditory stimuli were generated and controlled using Matlab, Psychtoolbox version 3.09 (revision 1754) (Brainard, 1997; Kleiner et al., 2007) running on an Apple MacBook Pro under Macintosh OS-X 10.6.7. Tones were digitized at a sampling rate of 44.8 kHz via the computer's internal HDA sound chip and presented binaurally via insert earphones (E-A-RTONE® 3A, Aero Company, USA). Precise on- and offsets of the stimuli were verified using photodiode and microphone recordings.

Experimental procedures

At the beginning of the experiment, participants' detection thresholds were measured for pure tones with a carrier frequency corresponding to the 5 target frequencies and a set of standard frequencies: 250, 500, 1000, 2000 and 4000 Hz. These subject-specific thresholds were used to scale the sounds' intensities separately for the masking and the target sounds to an equal level of loudness across different frequencies. Next, participants were familiarized with the stimuli and task in a total of 2-5 short test sessions of the informational masking paradigm. Only during those training sessions, did participants receive visual feedback after each trial.

The experiment comprised 8 sessions of 240 trials, separated on two days. After each session the participant's head position was adjusted to fit the position of the first session as accurately as possible (mean error ± standard deviation for day1: 4.8 ± 2.1 mm, day2: 4.4 ± 1.9 mm and across days: 7.5 ± 4.0 mm).

Data acquisition

Neuromagnetic data were recorded at 1171.88 Hz sampling frequency with a 275-channel whole-head MEG System (VSM, MedTech, Port Coquitlam, Canada; 275 axial gradiometers with 5 cm baseline and 29 reference channels) at the MEG Centre Tübingen, Germany. Participants' head position was continuously monitored by three sensor coils attached to the nasion, and left and right pre-auricular (15 mm anterior to the left and right tragus) points of each subject. The positions of these coils, i.e. the fiducial points, were marked on the subject's skin. To measure eye movements and blinks, horizontal and vertical electrooculogram (EOG) were recorded from two pairs of bipolar electrodes.

A 3T Siemens Magnetom Tim Trio System (Siemens, Erlangen, Germany) at the MPI for Biological Cybernetics, Tübingen, Germany, was used to acquire high-resolution structural images (176 sagittal slices, TR = 2300 ms, TE = 2.98 ms, TI = 1100 ms, flip angle = 9°, FOV = 256 mm x 240 mm x 176 mm, voxel size = 1 mm x 1 mm x 1 mm). MR-markers that can be identified on the anatomical image were attached to the same fiducial points as described above to enable accurate co-registration of the anatomical MRI and the MEG data.

Analysis

The MEG, MRI and behavioural data were pre-processed and analysed using statistical parametric mapping SPM8 (http://www.fil.ion.ucl.ac.uk/spm/; Wellcome Trust Centre of Neuroimaging, London, UK), fieldtrip (http://www.ru.nl/donders/fieldtrip) (Oostenveld et al., 2011) and Matlab 7 (Mathworks, Inc., Massachusetts, USA).

Sensor space: Pre-processing and statistics

For the sensor space analysis, we focused on event-related magnetic fields. Therefore, the continuous data were high-pass (cut-off: 1 Hz) and low-pass (cut-off: 30 Hz) filtered in forwards and reverse directions, using a 5th order Butterworth digital filter. Hence, very low-frequency drifts and the 40 Hz SSR were removed. The filtered data were then down-sampled to 213.3 Hz and epoched into segments from -170 ms to 880 ms after the onset of the target tone (adjusted for a visual delay of the projector of 20 ms). The epoched data were baseline corrected by subtracting the activity averaged between -150 to -50 ms from all MEG channels. Noisy epochs (i.e. 3.9 % of all epochs) were rejected when the MEG signal exceeded 1.8 pT. Hence, jump, muscle and other non-physiological artefacts were removed. Independent component analysis (ICA) was applied to correct for eye blink and heart beat artefacts. Eye blink and heartbeat-related components were identified based on visual inspection of component topographies and time-courses. In all datasets one single ICA component was related to eye blinks, while 1-2 (mean: 1.8) ICA components were related to heart beats. Finally, artefact-cleaned epochs were merged across sessions and averaged across trials to create event-related fields.

The linearly interpolated topography x time data were converted to 3D images (voxel size: 2.1 mm x 2.7 mm x 4.7 ms, image dimension: 64 x 64 x 214). The resulting images were smoothed in space and time, using an isotropic Gaussian Kernel of 12 mm/ms full-width at half maximum. At the random effects or between-subject level, for each subject 3D images were entered into several paired t-tests. Specifically, we performed the three statistical comparisons:

(1) The effect of *awareness* on processing the first auditory target: Hit1 vs. Miss1.

(2) The effect of *awareness* on processing the second auditory target: Hit2 vs. Miss2.

(3) The interaction between *awareness* and *target-position*: Hit1-Miss1 vs. Hit2-Miss2.

Based on our a priori hypotheses we restricted the analysis to the three time windows of interest (1) the early M1 response (40-90 ms) (Boutros and Belger, 1999; Wiegand and Gutschalk, 2012), (2) the middle latency ARN (100-200ms) (Gutschalk et al., 2008; Konigs and Gutschalk, 2012; Wiegand and Gutschalk, 2012) and (3) the later long-latency M3 (250-550 ms) (Ishii et al., 2009). Time windows' latencies were selected *a priori*, guided by previous studies and visual inspection of the mean time courses, pooled over all conditions and participants (critically, this does not invalidate our inference as the mean activity across all conditions is orthogonal to our contrasts of interest). Unless otherwise stated, we report effects at p < 0.05 at the peak level corrected for multiple comparisons, within the entire interpolated scalp space and the time windows of interest, using random field theory.

Source space analysis: MRI processing, MEG-MRI coregistration and forward modelling

Structural MRI images were segmented and normalized to MNI space using unified segmentation (Ashburner and Friston, 2005). The inverse of this normalization transformation was employed to warp a template cortical mesh, i.e. a continuous tessellation of the cortex (excluding cerebellum) with 20484 vertices, from MNI space to each subject's native space. The MEG data were projected onto each subject's MRI space by applying a rigid body coregistration using the fiducials as landmarks. As head model, we employed a single shell aligned with the inner skull. Lead fields were then computed for each vertex in the cortical mesh with each dipole oriented normally to that mesh.

Model inversion and source space statistics

Source localization was performed within a Bayesian framework using the Greedy Search (GS) algorithm implemented in SPM8 (version r4667), individually for each participant within a time-window from 0 to 600 ms. This time-window includes all windows of interest from our sensor space analysis. For each participant, the bandpass filtered (1-30 Hz) MEG data for all conditions were convolved with a Hanning window (used to down-weight baseline noise) and inverted together using 1024 patches per hemisphere (plus 1024 bilateral patches) of the cortical mesh.

Before inversion, the data were projected to a subspace of 73-102 (across-subjects mean: 81.7) spatial modes, based on a singular value decomposition (SVD) of the outer-product of the leadfield matrix, to retain > 92.8 % of the data variance. The projected data were further reduced to approximately 14-23 temporal modes (across-subjects mean: 19.7; with the maximum number of temporal modes set to 32) based on the SVD of the data variance matrix.

Additionally, anatomical 'soft' priors (Litvak et al., 2011) were used that were defined by the AAL library (Tzourio-Mazoyer et al., 2002) using the MarsBaR toolbox (http://marsbar.sourceforge.net/) (Brett et al., 2002): Bilateral Heschl's gyri; Bilateral superior and inferior parietal cortex and bilateral superior and middle frontal gyri.

The inversion scheme calculated source time-courses at each vertex in the cortical mesh for each condition and subject. For statistical analysis, the average energy of the source time-course were computed over the entire 600 ms window for all frequencies between 2 and 30 Hz; the source energies were then interpolated into volumetric images in MNI space with 2 mm voxels and spatially smoothed with a 12 mm FWHM isotropic Gaussian kernel. At the random effects or between-subject level, one source energy image per condition and subject was entered into paired t-tests. As in sensor space, we performed the following three statistical comparisons:

(1) The effect of *awareness* on processing the first auditory target: Hit1 vs. Miss1.

(2) The effect of *awareness* on processing the second auditory target: Hit2 vs. Miss2.

(3) The interaction between *awareness* and *target-position*: Hit1-Miss1 vs. Hit2-Miss2.

To obtain coordinates for later source wave extraction, we additionally calculated the main effect of *awareness*: (Hit1 + Hit2) vs. (Miss1 + Miss2). The results were used for source wave extraction (see below).

Extracting source waveforms

Source waveforms were extracted as the first eigenvariate of all vertices being significantly different for hits relative to misses as determined by the main effect of awareness (± 2 mm) (**Figure 3A**). As the polarity of the first eigenvariate is not uniquely defined for MEG data, we determined (i.e. changed) the polarity of the source time courses for each subject such that the consistency across participants was maximized (based on correlation and a 2nd order singular value decomposition of the time courses across all participants). The ensuing source time-courses were used for regionally-selective analyses of the event related responses and the steady-state responses.

Steady-State responses (SSRs): Analysis and statistics

Steady-state responses were characterized only in source space, because the brief amplitude modulated tones elicited only weak steady-state activity that may evade sensor space analyses as a consequence of intersubject variability in neuroanatomy, positioning of the subject etc.. Furthermore, it is well established that auditory steady-state activity is generated in auditory cortices enabling us to impose priors on the source localization (Schoonhoven et al., 2003; Millman et al., 2010; Okamoto et al., 2010; Steinmann and Gutschalk, 2011; Giani et al., 2012).

To optimize model inversion selectively, we pre-processed the data differently for later characterization of steady-state responses in source space. In particular, the pre-processing was identical to the one described for event related responses, except that the continuous data were low-pass filtered with a cut off at 100 Hz and epoched into segments from 0 ms to 300 ms (i.e. 300 ms length, including 12 cycles of the 40 Hz steady-state activity). Accordingly, the temporal window during source inversion was adjusted to range from 0-300 ms. Moreover, since SSR have been shown to be localised within the auditory cortex, only bilateral

Heschl's Gyrus was used as a spatial prior. Otherwise the inversion was identical to the one described for even related responses.

Before inversion, the data were projected to a subspace of 72-101 (across-subjects mean: 81.8) spatial modes, based on a singular value decomposition (SVD) of the outer-product of the leadfield matrix, to retain 95.2 % of the data variance. The projected data were further reduced to approximately 14-24 (across-subjects mean: 18.75) temporal modes (across subjects-mean) based on the SVD of the data variance matrix.

We extracted source-waveforms from bilateral auditory ROIs as described above. To estimate the amplitude at 40 Hz, we applied a fast Fourier Transform to the extracted time courses. First, we addressed the question if our stimuli evoked reliable SSRs. Therefore, we compared the amplitude at 40 Hz to the pooled amplitude of the adjacent sideband frequency. In paired t-tests we asked if amplitude at 40 Hz was significantly higher than at the sideband frequencies. Next, we estimated the effect of awareness by subtracting the amplitude of hits from the amplitude of misses. In paired t-test we evaluated whether the effect at 40 Hz was significantly higher than the effect at sideband frequencies, separately for each target with the sequence.

Effective Connectivity Analysis: DCM

Using dynamic causal modelling (DCM) we asked which connections in the auditory-parietal circuitry are modulated by awareness. Specifically, we hypothesised that early awareness effects, such as the ARN, are caused by modulations of the intrinsic connections within auditory cortices. In contrast, later awareness effects might rather be caused by modulations in extrinsic (i.e. feedforward and feedback) connections between auditory and parietal cortices.

Since awareness effects were most prominent for the second target, we specifically modelled awareness effects evoked by the second target within a sequence (i.e. Hit2 relative to Miss2), using DCM. To assess early changes in effectively connectivity we modelled time courses only from 0 – 300 ms, for later changes we looked at the whole temporal dynamics ranging from 0 – 550 ms.

For each participant, 6 DCMs were constructed. These models did not differ in terms of their nodes and connections: Subcortical, auditory input was assumed to arrive at the bilateral auditory cortices, which were connected laterally. Right parietal cortex was connected reciprocally to bilateral auditory cortices, i.e. through feedforward and feedback connections (**Figure 4A**). Holding the anatomical connections constant, differences between the models were based exclusively on the number and type of connections that were allowed to show awareness dependent influences on coupling strength. Extrinsically, we allowed awareness to influence coupling strength through (1) feedforward (**Figure 4A, red arrows**), (2) feedback (**Figure 4A, purple arrows**) or (3) reciprocal connections, i.e. through feedforward and feedback. Furthermore, intrinsic modulations within auditory cortices were either present or absent. Our 6 models can be summarized in a 3 x 2 factorial model space, modulating either extrinsic connections, i.e. feedforward (FF), feedback (FB) or both (FF /FB), or intrinsic connections, i.e. present (i+) or absent (i-).

Using the DCM for evoked responses we modelled the sourcewaves extracted previously. Hence, no spatial model was selected (option: 'LFP'). Before the analysis sourcewaves were normalized across conditions, because DCM assumes roughly equivalent scaling. No further processing (such as detrending) was applied. Afferent auditory input was parameterized as a Gaussian. The input was assumed (i.e. using soft priors) to activate auditory cortices around 100 ms (standard deviation: 50).

To determine the most likely model given the data, the 6 models were compared using random effect Bayesian model selection separately for the early and late time windows. To characterize our Bayesian model selection results we report the expected probability of one model being more likely than any other model tested.

RESULTS

Behaviour

Participants detected on average 39.67 ± 10% (mean ± standard deviation) and had 7.08 ± 4.73% false alarms, resulting in an across participants mean d' (i.e. d prime) of 1.32 ± 0.53 (across participants mean ± standard deviation). As previous studies suggested that the detection rate depends on target frequency (Bregman, 1990; Gutschalk et al., 2008), we also evaluated the detection rate separately across target frequencies. Indeed, a repeated measures ANOVA identified a significant effect of *target-frequency* on the detection rate (F (1.93) = 15.77, p < 0.001). Post hoc tests showed that the targets with the lowest target frequency (i.e.1222 Hz) were less often detected than all the other targets. Likewise, targets with a frequency of 1583 Hz were less often detected than targets at 2049 Hz (**table 1**).

Table 1. Behavioural results: Detection rates. Mean hit rates and correct rejections (± standard deviation) are shown separately for the 5 different target frequencies.

	Target frequency				
	1222 Hz	**1583 Hz**	**2049 Hz**	**2654 Hz**	**3437 Hz**
Hit (%)	15.35 ± 7.50	35.07± 17.25	51.98± 23.18	48.76± 19.21	45.92 ± 8.88
Corr.Rej. (%)	90.55 ± 6.65	90.79 ± 7.31	91.53 ± 6.05	92.79 ± 6.70	93.44 ± 5.42

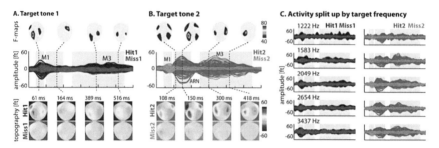

Figure 2. Sensor space results. (*A*) Butterfly plot of the activity evoked by the first target tone. Significant differences between hit and miss conditions are shown in time (dashed line). Statistical differences in space are shown as F-maps (top) and topographies (bottom). (*B*) The same plot as in (A) is shown for the second target tone. (*C*) Butterfly plots of the evoked activities for the first (left column) and second (right column) target tones, are shown for each target frequency separately.

Sensor space

In sensor space we investigated the effect of awareness on auditory processing by comparing the activity for detected and undetected auditory targets in paired *t*-tests separately for target 1 and 2. Furthermore, we directly tested for the interaction between *awareness* and *target-position*. Based on a priori hypotheses we evaluated the effect of awareness in three time-windows of interest: (1) M1: 40-90 ms, (2) ARN: 100-200ms, (3) M3: 250-550 ms (**Figure 2, table 2**).

A neuronal processing cascade
Irrespective of participants' awareness all auditory targets evoked an early M1 response with a bipolar topography over bilateral temporal sensors, i.e. probably the magnetic equivalent to the P50 (Woldorff et al., 1993; Woldorff et al., 1998; Boutros and Belger, 1999). However, for the first target tone only, we observed a significant difference in the M1 component for hits and misses at ~60 ms, indicating that awareness affects even very early auditory processing

(**Figure 2A**). Further, we identified a significant *awareness* x *target-position* interaction at 85 ms suggesting that the effects of awareness also depends on the target's sequential position. In the mid-latency window, we observed a significant difference in neural activity for detected relative to undetected auditory targets selectively when they come second (**Figure 2B**). When detected, the second target elicited an additional deflection with a dipolar topography opposite to the polarity of the M1 component. In line with the previously reported awareness related negativity (ARN) (Gutschalk et al., 2008; Konigs and Gutschalk, 2012; Wiegand and Gutschalk, 2012), the activity emerged at approximately 100 ms, peaked at ~ 150 ms and ended at ~200ms. Critically, as statistically confirmed by a significant *awareness* x *target-position* interaction, this awareness related negativity was evident only for the second target, but not for the first target. This activity profile suggests that segmentation and perception of the sequential pair of tones induces the ARN.

Detected relative to undetected targets evoked increased M3 deflections (250 to 550 ms) (Bledowski et al., 2006; Ishii et al., 2009). This awareness related effect was identified for both target 1 and target 2 (**Figure 2 A & B**). More specifically, for target activity differences for hit and misses were observed from 380 to 510 ms. For target 2, they peaked around 300 and 418 ms. No significant *awareness* x *target-position* interaction was detected.

Evaluation of the influence of target frequency on the sensor space results

At the behavioural level, the target detection rate significantly depended on carrier frequency of auditory target. This raises the question, whether the observed effects of awareness may in fact reflect effects induced by different carrier frequencies. To evaluate this potential confound, we reanalyzed our data by sorting hits and misses into 5 new conditions according to the target's carrier frequency. At the random effects level, we then entered these data into a 2 (awareness: hits, misses) x 2 (target position: first, second) x 5 (carrier frequency: 1222, 1583, 2049, 2654 and 3437 Hz) repeated measure ANOVA. This ANOVA identified no significant effects at $p < 0.05$ (corrected for multiple comparisons). Likewise, **figure 3C** shows that the effect of awareness was similar for all target frequencies.

Table 2. Statistical results of the sensor space analysis. Statistical comparisons and results are shown at p < 0.05 corrected. Image coordinates (mm*mm*ms) are given for the peak voxel. Sensor locations correspond to CTF conventions (first letter: R = right and L = left; second letter: T = temporal, O = occipital, P = parietal, F = frontal; numbering: 1st digit = row and 2nd digit = column). P-values are corrected for multiple comparisons for all voxels at time windows of interest.

Coordinates			Nearest	Z-score	p-value
mm	mm	ms	sensor	(peak)	(corrected)
Effect of awareness					
Hit1 ≠ Miss 1					
-32	-38	61	LT16	5.65	< 0.001
30	16	164	RF55	4.98	0.007
28	-17	389	RC25	4.74	0.018
-30	-33	516	LP55	4.54	0.039
Hit2 ≠ Miss 2					
-30	-46	150	LT27	5.45	0.001
-17	-17	418	LP23	5.19	0.002
55	-17	300	RT34	5.06	0.004
-53	21	160	LT33	4.75	0.014
36	-38	108	RT16	4.52	0.035
Awareness x target-position (interaction)					
(Hit1 – Miss1) ≠ (Hit2 – Miss2)					
-34	-38	155	MLT16	5.36	0.001
-38	-46	85	MLT37	4.99	0.006

Source space analysis

3.2.1 Event related responses

Within source space we localized the effect of awareness (**Table 3, Figure 3**). Hits relative to misses induced significant increases in source energy within bilateral auditory cortices (i.e. Heschl's Gyrus) for both target tones. (However, right HG is significant only at 0.001 uncorrected for the second target). At a more liberal threshold of 0.001 (uncorrected), right parietal activation was seen for both targets. Accordingly, the main effect of awareness showed significantly enhanced source energy for hits relative to misses in bilateral auditory cortices and right parietal cortex (**Figure 3A**). Lastly, a significant *awareness* X *target-position* interaction was observed in left auditory cortex.

Figure 3B shows 1 source-wave per condition, extract from bilateral auditory and right parietal cortices. It appears that all three components of interest were localized within the auditory cortex. Late M3-like activity was additionally localized to the right parietal ROI.

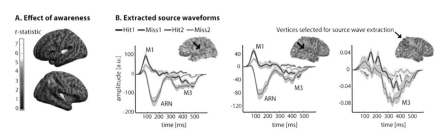

Figure 3. Source space results. (*A*) Main effect of awareness. Statistical results are displayed on a canonical, inflated MNI brain. (*B*) Extracted source waves. Source waves of hit1 (dark blue), hit2 (light blue), Miss1 (red) and Miss2 (oranges) are displayed along with shaded error bars for the left auditory ROI (left column) right auditory ROI (middle column) and parietal ROI (right column). Vertices that are selected for source wave extraction are represented as red dots on a representative participant's cortical mesh.

Table 3. Statistical results of the source space analysis. Statistical comparisons and results are shown at p < 0.05 corrected at the peak level for multiple comparisons within search volumes (i.e. left and right Heschl's Gyrus, left and right superior and inferior parietal cortex, superior and middle frontal gyri).

Brain Region	MNI Coordinates			Cluster size	Z-score (peak)	p-Value (corrected)
	x	y	z			
Effect of awareness						
Hit1 > Miss1						
R. transverse temporal gyrus (HG)	50	-12	4	155	4.54	0.016
L. transverse temporal gyrus (HG)	-46	-18	6	124	4.53	0.017
R. inferior parietal cortex (supramarginal gyrus)	48	-46	50	982	3.69	0.264
Hit2 > Miss 2						
L. transverse temporal gyrus (HG)	-42	-20	6	123	4.77	0.004
R. transverse temporal gyrus (HG)	48	-16	6	138	3.98	0.079
R. intraparietal sulcus	46	-46	50	91	3.30	0.479
R. transverse parietal sulcus	30	-56	50	37	3.26	0.512
R. superior parietal gyrus	40	-46	58	15	3.18	0.593
R. intraparietal sulcus	34	-40	52	1	3.10	0.985
Awareness x target-position interaction						
(Hit1 – Miss1) ≠ (Hit2 – Miss2)						
L. transverse temporal gyrus (HG)	-44	-20	6	93	3.40	0.109
Main effect of awareness						
(Hit1 + Hit2) ≠ (Miss1 + Miss2)						
L. transverse temporal gyrus (HG)	-44	-20	6	127	5.07	< 0.001
R. transverse temporal gyrus (HG)	46	-16	6	151	4.45	0.002
R. inferior parietal cortex (supramarginal gyrus)	48	-46	50	1105	3.85	0.022

Steady-State Responses

As expected, the 40 Hz amplitude modulated target tone evoked a steady-state response that closely followed the frequency of stimulation. Correspondingly, the amplitude at 40 Hz proved to be significantly increased relative to the sideband frequencies' amplitude. Hence, target tones elicited reliable SSRs.

More importantly however, we were interested in how this SSR was affected by awareness. Generally detected targets appeared to elicit stronger SSRs than misses. However, this increase in amplitude is not specific to 40 Hz frequency. In fact, the effect of awareness did not diverge significantly from the effect at sideband frequencies (**table 4**).

Table 4. Statistical comparisons and results of the steady-state responses. Statistical comparisons and results of the analysis of steady-state responses. (HG = Heschl's Gyrus)

	Target 1		Target 2	
	Hit 1	*Miss 2*	*Hit2*	*Miss2*
amplitude 40 Hz ≠ sideband frequency				
Left HG	$t(19) = 2.70$, $p = 0.014$	$t(19) = 4.80$, $p < 0.001$	$t(19) = 2.82$, $p < 0.011$	$t(19) = 4.30$, $p < 0.001$
Right HG	$t(19) = 4.15$, $p < 0.001$	$t(19) = 5.35$, $p < 0.001$	$t(19) = 4.47$, $p < 0.001$	$t(19) = 4.72$, $p < 0.001$
Hit – Miss: amplitude 40 Hz ≠ sideband frequency				
Left HG		$t(19) = -0.19$, $p = 0.852$		$t(19) = 0.65$, $p = 0.525$
Right HG		$t(19) = 0.20$, $p = 0.847$		$t(19) = 1.16$, $p = 0.262$

Effective connectivity: DCM

Using DCM we tried to model source-waves, reconstructed previously. In doing so, we were interested how modulations in coupling strength within and between sources influenced these source-waves.

We compared 6 models, differing exclusively in the number and type of connections that were allowed to show awareness dependent changes in coupling strength. **Figure 4B** shows the expected probabilities of each model in our factorial 2 x 3 model space. Generally, it can be seen that models comprising awareness dependent intrinsic modulations in bilateral auditory cortex (i.e. i+ models) outperform models missing these awareness dependent modulations. This finding was independent from the modelled epoch length (i.e. short/brief or long).

However, while the feedforward (FFi+) model best explains the first 300 ms of the data, the feedback model (FBi+) outperforms the other models in explaining the complete dynamics (i.e. 500 ms). **Figure 4C** presents the modelled and observed time courses for the best model (i.e. feedforward for the early and feedback for the late epochs), using the grand-mean response over participants. The model accurately models the ARN component. Long latency effects are likewise modelled, though less accurately.

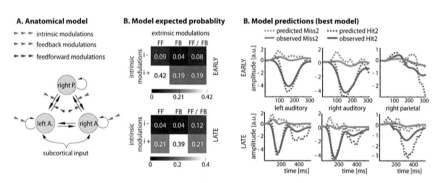

Figure 4. Dynamic causal modelling (DCM). (*A*) Anatomical DCM. From this basic model six candidate DCMs were generated by factorially modulating (1) extrinsic connections (i.e. forward, feedback or both) and (2) intrinsic connections within the auditory cortex. (*B*) Bayesian model comparison (random effects analysis) for early (< 300 ms, upper row) and late (0 – 550 ms, lower row) dynamics. The matrix shows the expected probability of the six DCMs in a factorial fashion. (*C*) Mean predicated source activity estimates (dashed line) and observed source activity (continuous line) of the winning model for hit2 (blue) and miss2 (orange) conditions. Upper row shows source estimates for the early temporal dynamics, while the lower row depicts the late temporal dynamics.

DISCUSSION

Using informational masking and MEG we identified a neuronal processing cascade that enabled awareness of a pair of target tones, embedded within a multi-tone background. Awareness selectively enhanced early processing of the first target. Yet, only after onset of the second target, both targets could be integrated into a coherent stream. Following this integration an ARN component occurred which crucially depended on intrinsic modulations within the auditory cortex. Lastly, both target tones evoked a sustained, late M3 component, which relied on recurrent activation within a temporo-parietal network. It probably reflects working memory operations and perceptual awareness.

A neuronal processing cascade

Independent of participants' perceptual experience, the first target tone evoked a M1 component around 60 ms. Yet, the component's amplitude was increased for detected relative to undetected tones, pointing to a gating mechanism that selectively enhanced relevant sounds. In line, increased evoked activity around 60 ms was shown to reflect the system's capacity to "gate in" relevant information. Irrelevant information instead is filtered out to prevent the system from overload (Boutros and Belger, 1999; reviewed in: Patterson et al., 2008). Awareness thus increases processing of the first target.

Critically however, the tasks required participants to detect a *sequence* of two target tones. They thus had two temporally integrate a pair of targets into a coherent stream. This temporal integration could only be completed after onset of the second target tone. Interestingly, only after onset of this very target tone we observed a robust, ARN-like deflection. The ARN thus correlated with the conscious formation of a target stream. This result challenges previous findings, which suggested that the ARN parallels the detection of *each* target tone within a sequence (Gutschalk et al., 2008).

Following both target tones we observed a sustained deflection (i.e. the M3 component). An analogous long-latency component (called P3 or P3b in EEG) was previously associated with

101

working memory processes that are necessary to evaluate a stimulus (Kok, 2001; Bledowski et al., 2006). Interestingly, different stages in working memory processing, may be paralleled by multiple subcomponents. The M3 components described here peaked twice: First between 300-400 ms and second between 400-500 ms. While, evaluating a stimulus may evoke early subcomponents, later subcomponents may rather reflect storage buffer operations (Bledowski et al., 2006).

Previous research moreover showed that increasing memory load decreases the amplitude while increasing the latency of the P3 component (Kok, 2001; Bledowski et al., 2006). Even though there was no significant awareness x target position interaction, **figure 2** suggests that the first relative to the second detected target evokes a lower amplitude at a longer latency. We speculate that this result reflects the increased difficulty of stimulus evaluation following the first target. The P3 component is also associated with perceptual awareness and attention (e.g.: Hillyard et al., 1971; Kok, 2001; Sergent et al., 2005; van Aalderen-Smeets et al., 2006; Del Cul et al., 2007; Bekinschtein et al., 2009; Ishii et al., 2009; Shen and Alain, 2011; Auksztulewicz et al., 2012). Therefore, it may also reflect perceptual awareness of each target tone or attentional shifts.

Collectively, our results suggest that awareness selectively enhances processing of the first target tone. Only after the onset of the second target both target tones can be integrated into a coherent stream. This process is paralleled by an ARN. Lastly, a sustained M3 component reflects working memory operations and probably perceptual awareness.

Reconstructing source-waves

Along with previous informational masking studies, our source space analysis showed that awareness effects neuronal processing within the auditory cortex (Gutschalk et al., 2008; Konigs and Gutschalk, 2012; Wiegand and Gutschalk, 2012). Yet, extending these results, we found an additional M3 component to be localized in parietal cortices. In line, sustained parietal activation was associated with awareness and working memory (Sergent et al., 2005;

Bledowski et al., 2006; Dehaene et al., 2006; Del Cul et al., 2007; Bekinschtein et al., 2009; Shen and Alain, 2011).

Moreover, running a DCM analysis we could examine the effective connectivity within the auditory-parietal network. In particular, we found that intrinsic modulations within the auditory cortex are crucial to model the early (i.e. < 300 ms) temporal dynamics (i.e. particularly the ARN component). We assume that these modulations are necessary to match and to integrate both target tones. This interpretation is support by a previous DCM study that showed the importance of intrinsic auditory modulations to process sound deviants (Garrido et al., 2009).

While early dynamics mainly relied on intrinsic, auditory modulations, later (> 300 ms) activation rather depended on extrinsic modulations of recurrent connections between auditory and parietal cortices. In line, there is a vast amount of literature showing that M3-like components are generated within a broad temporo-parieto-frontal network (Dehaene et al., 2006). Similarly, fMRI based evidence suggests that perceptual awareness correlates with activations in parietal and frontal cortices (Beck et al., 2001). Lastly, a previous DCM study showed that recurrent neuronal processing is substantial for somatosensory awareness (Auksztulewicz et al., 2012). Along with the global network hypothesis (Dehaene et al., 2006), we therefore conclude that awareness of a sequence of target tones depends on top-down influences of higher order parietal cortices.

Yet, in contrast to our hypothesis, we could not find any awareness related effects within frontal cortices. We assume that, large anatomical variability of frontal sources across participants, as well as less precise anatomical priors, complicated source analysis of frontal relative to auditory sources and concealed awareness effects.

Difference to previous studies

Generally our data nicely fit with most studies discussed above. However, challenging previous results, we found early awareness related effects (Gutschalk et al., 2008; Konigs and Gutschalk, 2012; Wiegand and Gutschalk, 2012). We assume that the apparently contradictory results were found, because we did not average evoked activity over multiple target tones. The ARN, which occurs after onset of the second target, partially overlaps with the M1 component. Averaging MEG activity evoked by multiple targets may thus conceal early awareness related effects. Interestingly, in at least one study evoked responses of individual targets are shown (Wiegand and Gutschalk, 2012). Even though this finding is not discussed it appears that the activity ~60 ms is enhanced for the first detected target.

Surprisingly, using informational masking previous MEG studies did not report on any M3-like deflections. Yet, these later, top-down modulations of fronto-parietal networks seem crucial for perceptual awareness. Since authors restricted analysis to auditory cortices they might have missed these recurrent activations from higher-order cortices. Moreover it may be that the M3 gets attenuated during a sequence comprising various tones.

REFERENCES

Ashburner J, Friston KJ (2005) Unified segmentation. Neuroimage 26:839-851.

Auksztulewicz R, Spitzer B, Blankenburg F (2012) Recurrent neural processing and somatosensory awareness. J Neurosci 32:799-805.

Beck DM, Rees G, Frith CD, Lavie N (2001) Neural correlates of change detection and change blindness. Nat Neurosci 4:645-650.

Bekinschtein TA, Dehaene S, Rohaut B, Tadel F, Cohen L, Naccache L (2009) Neural signature of the conscious processing of auditory regularities. Proc Natl Acad Sci U S A 106:1672-1677.

Bledowski C, Cohen Kadosh K, Wibral M, Rahm B, Bittner RA, Hoechstetter K, Scherg M, Maurer K, Goebel R, Linden DE (2006) Mental chronometry of working memory retrieval: a combined functional magnetic resonance imaging and event-related potentials approach. J Neurosci 26:821-829.

Boutros NN, Belger A (1999) Midlatency evoked potentials attenuation and augmentation reflect different aspects of sensory gating. Biol Psychiatry 45:917-922.

Brainard DH (1997) The Psychophysics Toolbox. Spat Vis 10:433-436.

Bregman AS (1990) Auditory scene analysis: The perceptual organization of sound. Cambridge, MA: MIT press.

Brett M, Anton J, Valabregue R, Poline JB (2002) Region of interest analysis using an SPM toolbox. Neuroimage 16:Presented at the 8th International Conference on Functional Mapping of the Human Brain, June 2-6, 2002, Sendai, Japan.

Cusack R (2005) The intraparietal sulcus and perceptual organization. J Cogn Neurosci 17:641-651.

Cusack R, Deeks J, Aikman G, Carlyon RP (2004) Effects of location, frequency region, and time course of selective attention on auditory scene analysis. J Exp Psychol Hum Percept Perform 30:643-656.

Dehaene S, Changeux JP, Naccache L, Sackur J, Sergent C (2006) Conscious, preconscious, and subliminal processing: a testable taxonomy. Trends Cogn Sci 10:204-211.

Del Cul A, Baillet S, Dehaene S (2007) Brain dynamics underlying the nonlinear threshold for access to consciousness. PLoS Biol 5:e260.

Durlach NI, Mason CR, Kidd G, Jr., Arbogast TL, Colburn HS, Shinn-Cunningham BG (2003) Note on informational masking. J Acoust Soc Am 113:2984-2987.

Garrido MI, Kilner JM, Kiebel SJ, Friston KJ (2009) Dynamic causal modeling of the response to frequency deviants. J Neurophysiol 101:2620-2631.

Giani AS, Ortiz E, Belardinelli P, Kleiner M, Preissl H, Noppeney U (2012) Steady-state responses in MEG demonstrate information integration within but not across the auditory and visual senses. Neuroimage 60:1478-1489.

Gutschalk A, Micheyl C, Oxenham AJ (2008) Neural correlates of auditory perceptual awareness under informational masking. PLoS Biol 6:e138.

Hillyard SA, Squires KC, Bauer JW, Lindsay PH (1971) Evoked potential correlates of auditory signal detection. Science 172:1357-1360.

Ishii R, Canuet L, Herdman A, Gunji A, Iwase M, Takahashi H, Nakahachi T, Hirata M, Robinson SE, Pantev C, Takeda M (2009) Cortical oscillatory power changes during auditory oddball task revealed by spatially filtered magnetoencephalography. Clin Neurophysiol 120:497-504.

Kleiner M, Brainard D, Pelli D (2007) What's new in Psychtoolbox-3? Perception 36.

Kok A (2001) On the utility of P3 amplitude as a measure of processing capacity. Psychophysiology 38:557-577.

Konigs L, Gutschalk A (2012) Functional lateralization in auditory cortex under informational masking and in silence. Eur J Neurosci.

Lamme VA, Roelfsema PR (2000) The distinct modes of vision offered by feedforward and recurrent processing. Trends Neurosci 23:571-579.
Litvak V, Mattout J, Kiebel S, Phillips C, Henson R, Kilner J, Barnes G, Oostenveld R, Daunizeau J, Flandin G, Penny W, Friston K (2011) EEG and MEG data analysis in SPM8. Comput Intell Neurosci 2011:852961.
Millman RE, Prendergast G, Kitterick PT, Woods WP, Green GG (2010) Spatiotemporal reconstruction of the auditory steady-state response to frequency modulation using magnetoencephalography. Neuroimage 49:745-758.
Moore CJM (1995) Frequency analysis and masking. In: Hearing (Moore CJM, ed), pp 161-205. USA: Academic Press, Inc.
Neff DL, Green DM (1987) Masking produced by spectral uncertainty with multicomponent maskers. Percept Psychophys 41:409-415.
Okamoto H, Stracke H, Bermudez P, Pantev C (2010) Sound Processing Hierarchy within Human Auditory Cortex. J Cogn Neurosci.
Oostenveld R, Fries P, Maris E, Schoffelen JM (2011) FieldTrip: Open source software for advanced analysis of MEG, EEG, and invasive electrophysiological data. Comput Intell Neurosci 2011:156869.
Patterson JV, Hetrick WP, Boutros NN, Jin Y, Sandman C, Stern H, Potkin S, Bunney WE, Jr. (2008) P50 sensory gating ratios in schizophrenics and controls: a review and data analysis. Psychiatry Res 158:226-247.
Picton TW, John MS, Dimitrijevic A, Purcell D (2003) Human auditory steady-state responses. Int J Audiol 42:177-219.
Ro T, Breitmeyer B, Burton P, Singhal NS, Lane D (2003) Feedback contributions to visual awareness in human occipital cortex. Curr Biol 13:1038-1041.
Schoonhoven R, Boden CJ, Verbunt JP, de Munck JC (2003) A whole head MEG study of the amplitude-modulation-following response: phase coherence, group delay and dipole source analysis. Clin Neurophysiol 114:2096-2106.
Sergent C, Baillet S, Dehaene S (2005) Timing of the brain events underlying access to consciousness during the attentional blink. Nat Neurosci 8:1391-1400.
Shen D, Alain C (2011) Temporal attention facilitates short-term consolidation during a rapid serial auditory presentation task. Exp Brain Res 215:285-292.
Steinmann I, Gutschalk A (2011) Potential fMRI correlates of 40-Hz phase locking in primary auditory cortex, thalamus and midbrain. Neuroimage 54:495-504.
Teki S, Chait M, Kumar S, von Kriegstein K, Griffiths TD (2011) Brain bases for auditory stimulus-driven figure-ground segregation. J Neurosci 31:164-171.
Tzourio-Mazoyer N, Landeau B, Papathanassiou D, Crivello F, Etard O, Delcroix N, Mazoyer B, Joliot M (2002) Automated anatomical labeling of activations in SPM using a macroscopic anatomical parcellation of the MNI MRI single-subject brain. Neuroimage 15:273-289.
van Aalderen-Smeets SI, Oostenveld R, Schwarzbach J (2006) Investigating neurophysiological correlates of metacontrast masking with magnetoencephalography. Advances in Cognitive Psychology 2:21-35.
Wiegand K, Gutschalk A (2012) Correlates of perceptual awareness in human primary auditory cortex revealed by an informational masking experiment. Neuroimage 61:62-69.
Woldorff MG, Hillyard SA, Gallen CC, Hampson SR, Bloom FE (1998) Magnetoencephalographic recordings demonstrate attentional modulation of mismatch-related neural activity in human auditory cortex. Psychophysiology 35:283-292.
Woldorff MG, Gallen CC, Hampson SA, Hillyard SA, Pantev C, Sobel D, Bloom FE (1993) Modulation of early sensory processing in human auditory cortex during auditory selective attention. Proc Natl Acad Sci U S A 90:8722-8726.

GENERAL DISCUSSION

Multisensory integration and awareness describe two key processes of human perception. They allow us to form a coherent, conscious representation of the external world. Nevertheless, in contrast to their impact, relatively little is known about their interrelation. It remains controversial, for instance, to what extent multisensory integration happens automatically or dependent on higher cognitive processes such as awareness (Talsma et al., 2010). Current conventional views assume that sensory information is integrated automatically, only if the stimulus complexity of the environment is low (Talsma et al., 2010). In **chapter 1** we therefore evaluated the automaticity of multisensory processes. Specifically, we investigated whether low complex, multisensory signals are integrated in the absence of awareness by combining spatial ventriloquism and continuous flash suppression. Challenging conventional views, we found that ventriloquism is profoundly modulated by perceptual awareness. Nevertheless, even unconsciously viewed flashes induced a small, but significant, ventriloquist effect. These results suggest that auditory and visual inputs are integrated into coherent spatial representations at both, pre- and post-aware processing stages. To further underpin this interpretation, future work should replicate these results while simultaneously measuring neural activity using M/EEG and /or fMRI.

Eventually, we were interested in the *neural dynamics* underlying conscious, auditory and visual perception. Following, we therefore measured neural (i.e. MEG) activity during auditory and visual perception. However, identifying multisensory integration remains a major challenge to neuroimaging (Goebel and van Atteveldt, 2009; James and Stevenson, 2012). In **chapter two**, we therefore reported whether steady-state responses (SSRs) can be used to unambiguously index signal integration within the human brain (Regan and Regan, 1988). Traditional approaches used transient stimuli, analysed in the time domain (e.g.: Giard and Peronnet, 1999; Molholm et al., 2002; Teder-Salejarvi et al., 2002; Murray et al., 2005; Talsma and Woldorff, 2005; Talsma et al., 2007). These approaches, however, cannot unequivocally index multisensory integration. Instead, crossmodulation frequencies denote an unequivocal test for non-linear interplay of multiple signals (Regan and Regan, 1988; Regan et al., 1995).

Hence, we hoped to find crossmodulation frequencies by studying Fourier spectra of auditory and visual steady-state responses. Finding these crossmodulation frequencies would open new research avenues for tracking the influence of cognitive processes in multisensory integration. For instance, many M/EEG studies showed that perceptual awareness can be tracked, combining steady-state responses and binocular rivalry (e.g.: Cosmelli et al., 2004; Kamphuisen et al., 2008; Sutoyo and Srinivasan, 2009). Along those lines, audiovisual crossmodulation frequencies could be used to study the influence of perceptual awareness on audiovisual integration.

Our results reliably identified crossmodulation frequencies **within** sensory modalities. However, **across** modalities no crossmodulation frequencies were found. In contrast, using traditional approaches, virtually all M/EEG studies found indications for multisensory interplay (e.g.: Giard and Peronnet, 1999; Molholm et al., 2002; Teder-Salejarvi et al., 2002; Murray et al., 2005; Talsma and Woldorff, 2005; Talsma et al., 2007). We therefore assume that steady-state responses are unsuitable to study multisensory processes because they lack common spatio-temporal dynamics, which cue coherent entities. Since SSRs seem unsuitable to study multisensory integration within the brain, we turned to traditional ways of studying multisensory integration and awareness in the following experiments.

Note however, that our design lacked behavioural evidence that could support the absence of signal integration. Therefore, it remains debatable whether the absence of crossmodulation effects can be ascribed entirely to the absence of multisensory integration. Theoretically, null results could be ascribed to an insensitivity of our analysis method. Future work may pursue this challenge by including behavioural tasks. For instance, bringing auditory and visual steady-state stimuli into spatial conflict may evoke ventriloquist-like perception *only* if signals are integrated. (To do so one has to trade off the illusions' strength with the SSR's amplitude, e.g. by choosing a stimulus length between 300-1000 ms.)

While we were ultimately interested in the neural mechanisms that enable multisensory awareness, we focused on auditory awareness in **chapter 3**. Combining MEG and informational masking assessed the neural processing cascade that enabled perceptual awareness of a pair of target tones. In particular, results showed that awareness entails complex neural dynamics that depend on the target's sequential position: Awareness of the target sequence selectively enhanced early processing of the first target, probably reflecting sensory gating mechanisms. In turn, the second target tone evoked an awareness related negativity (ARN), which paralleled the sequential integration of both target tones. This ARN relied on intrinsic modulation within the auditory cortex. Lastly, working memory processing was reflected in a sustained deflection peaking around 300 and 500 ms after the onset of each target tone.

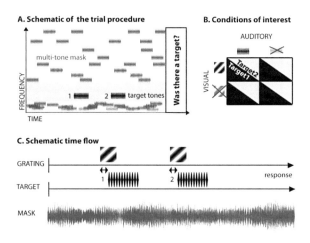

Figure 1: Experimental design and conditions. (*A*) The informational masking paradigm. Participants detected a pair of target tones (black) embedded within a multi-tone mask (grey). Target tones were presented in a protected region in frequency space (grey shading). After each trial participants indicated whether there was a target pair. (*B*) The 2 x 2 x 2 factorial design with the factors: (1) auditory target (present, absent), (2) visual target (present, absent) and (3) target position (first, second). (*C*) Schematic time flow.

Ultimately however, we were interested in the neural mechanisms that enable *multisensory* awareness and their relation to awareness. Therefore, additional trials comprised visual stimuli, too (figure 1). However, these trials were not included or discussed in chapter 3. Moreover, so far only preliminary results are available. These results show that awareness differently modulates multisensory integration early (target 1) and later (target 2) (Figure 2). In particular, we found an early difference around the visual M1 component (i.e. 150 ms) and immediately after the ARN at around 280 ms. These differences, may potentially index early pre-aware interactions that increase salience and facilitate auditory stream segregation. Future work is planned to further analyse these processes in sensor and source space.

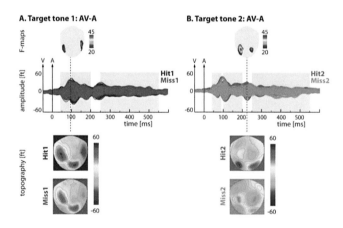

Figure 2: Sensor space results. (*A*) Butterfly plot of the activity evoked by the first target tone. Significant differences between hit1 and hit2 conditions are shown in time (dashed line). Statistical differences in space are shown as F-maps (top) and topographies (bottom). (*B*) The same information as in (A) is plotted for the second target tone.

Conclusion:

112

Our experiments investigated the (neural) mechanisms underlying audiovisual integration, perceptual awareness and their interrelation. Collectively our results suggest that:

(1) Auditory and visual inputs are integrated into coherent spatial representations at both, pre- and post-aware processing stages.

(2) Information represented in steady-state responses is integrated at the primary cortical level within but not across sensory modalities

(3) Awareness of a two-tone pair relies on a cascade of processes within a temporo-parietal network.

REFERENCES SYNOPSIS

Alais D, Burr D (2004) The ventriloquist effect results from near-optimal bimodal integration. Curr Biol 14:257-262.

Auksztulewicz R, Spitzer B, Blankenburg F (2012) Recurrent neural processing and somatosensory awareness. J Neurosci 32:799-805.

Battaglia PW, Di Luca M, Ernst MO, Schrater PR, Machulla T, Kersten D (2010) Within- and cross-modal distance information disambiguate visual size-change perception. PLoS Comput Biol 6:e1000697.

Beauchamp MS (2005) Statistical criteria in FMRI studies of multisensory integration. Neuroinformatics 3:93-113.

Beauchamp MS, Argall BD, Bodurka J, Duyn JH, Martin A (2004) Unraveling multisensory integration: patchy organization within human STS multisensory cortex. Nat Neurosci 7:1190-1192.

Bekinschtein TA, Dehaene S, Rohaut B, Tadel F, Cohen L, Naccache L (2009) Neural signature of the conscious processing of auditory regularities. Proc Natl Acad Sci U S A 106:1672-1677.

Bennett CM, Wolford GL, Miller MB (2009) The principled control of false positives in neuroimaging. Soc Cogn Affect Neurosci 4:417-422.

Bermant RI, Welch RB (1976) Effect of degree of separation of visual-auditory stimulus and eye position upon spatial interaction of vision and audition. Percept Mot Skills 42:487-493.

Bertelson P, Radeau M (1981) Cross-modal bias and perceptual fusion with auditory-visual spatial discordance. Percept Psychophys 29:578-584.

Bertelson P, Aschersleben G (1998) Automatic visual bias of perceived auditory location. Psychonomic Bulletin & Review 5:482-489.

Bertelson P, Vroomen J, de Gelder B, Driver J (2000a) The ventriloquist effect does not depend on the direction of deliberate visual attention. Percept Psychophys 62:321-332.

Bertelson P, Pavani F, Ladavas E, Vroomen J, de Gelder B (2000b) Ventriloquism in patients with unilateral visual neglect. Neuropsychologia 38:1634-1642.

Brancucci A, Franciotti R, D'Anselmo A, Della Penna S, Tommasi L (2011) The sound of consciousness: neural underpinnings of auditory perception. J Neurosci 31:16611-16618.

Bruce C, Desimone R, Gross CG (1981) Visual properties of neurons in a polysensory area in superior temporal sulcus of the macaque. J Neurophysiol 46:369-384.

Calvert GA (2001) Crossmodal processing in the human brain: insights from functional neuroimaging studies. Cereb Cortex 11:1110-1123.

Calvert GA, Hansen PC, Iversen SD, Brammer MJ (2001) Detection of audio-visual integration sites in humans by application of electrophysiological criteria to the BOLD effect. Neuroimage 14:427-438.

Calvert GA, Bullmore ET, Brammer MJ, Campbell R, Williams SC, McGuire PK, Woodruff PW, Iversen SD, David AS (1997) Activation of auditory cortex during silent lipreading. Science 276:593-596.

Chalmers DJ (1995) Facing up to the problem of consciousness. Journal of Consciousness Studies 2:200 - 219.

Chalmers DJ (2000) What is a neural correlate of consciousness. In: Neural correlates of consciousness: Empirical and conceptual questions (Metzinger T, ed), pp 17-39: MIT Press.

Cosmelli D, David O, Lachaux JP, Martinerie J, Garnero L, Renault B, Varela F (2004) Waves of consciousness: ongoing cortical patterns during binocular rivalry. Neuroimage 23:128-140.

Cowey A, Walsh V (2000) Magnetically induced phosphenes in sighted, blind and blindsighted observers. Neuroreport 11:3269-3273.

Dehaene S, Changeux JP (2011) Experimental and theoretical approaches to conscious processing. Neuron 70:200-227.

Dehaene S, Changeux JP, Naccache L, Sackur J, Sergent C (2006) Conscious, preconscious, and subliminal processing: a testable taxonomy. Trends Cogn Sci 10:204-211.

Del Cul A, Baillet S, Dehaene S (2007) Brain dynamics underlying the nonlinear threshold for access to consciousness. PLoS Biol 5:e260.

Diederich A, Colonius H (2004) Bimodal and trimodal multisensory enhancement: effects of stimulus onset and intensity on reaction time. Percept Psychophys 66:1388-1404.

Dixon NF, Spitz L (1980) The detection of auditory visual desynchrony. Perception 9:719-721.

Doesburg SM, Green JJ, McDonald JJ, Ward LM (2009) Rhythms of consciousness: binocular rivalry reveals large-scale oscillatory network dynamics mediating visual perception. PLoS One 4:e6142.

Driver J, Noesselt T (2008) Multisensory interplay reveals crossmodal influences on 'sensory-specific' brain regions, neural responses, and judgments. Neuron 57:11-23.

Durlach NI, Mason CR, Kidd G, Jr., Arbogast TL, Colburn HS, Shinn-Cunningham BG (2003) Note on informational masking. J Acoust Soc Am 113:2984-2987.

Engel AK, Singer W (2001) Temporal binding and the neural correlates of sensory awareness. Trends Cogn Sci 5:16-25.

Ernst MO, Bulthoff HH (2004) Merging the senses into a robust percept. Trends Cogn Sci 8:162-169.

Falchier A, Clavagnier S, Barone P, Kennedy H (2002) Anatomical evidence of multimodal integration in primate striate cortex. J Neurosci 22:5749-5759.

Fort A, Delpuech C, Pernier J, Giard MH (2002) Dynamics of cortico-subcortical cross-modal operations involved in audio-visual object detection in humans. Cereb Cortex 12:1031-1039.

Fries P, Roelfsema PR, Engel AK, Konig P, Singer W (1997) Synchronization of oscillatory responses in visual cortex correlates with perception in interocular rivalry. Proc Natl Acad Sci U S A 94:12699-12704.

Ghazanfar AA, Schroeder CE (2006) Is neocortex essentially multisensory? Trends Cogn Sci 10:278-285.

Giard MH, Peronnet F (1999) Auditory-visual integration during multimodal object recognition in humans: a behavioral and electrophysiological study. J Cogn Neurosci 11:473-490.

Goebel R, van Atteveldt N (2009) Multisensory functional magnetic resonance imaging: a future perspective. Exp Brain Res 198:153-164.

Gutschalk A, Micheyl C, Oxenham AJ (2008) Neural correlates of auditory perceptual awareness under informational masking. PLoS Biol 6:e138.

Haynes JD, Driver J, Rees G (2005) Visibility reflects dynamic changes of effective connectivity between V1 and fusiform cortex. Neuron 46:811-821.

Hettinger LJ, Berbaum KS, Kennedy RS, Dunlap WP, Nolan MD (1990) Vection and simulator sickness. Mil Psychol 2:171-181.

Hillebrand A, Barnes GR (2002) A quantitative assessment of the sensitivity of whole-head MEG to activity in the adult human cortex. Neuroimage 16:638-650.

James TW, Stevenson RA (2012) The Use of fMRI to Assess Multisensory Integration. In: The Neural Bases of Multisensory Processes (Murray MM, Wallace MT, eds). Boca Raton (FL): CRC Press.

Jousmaki V, Hari R (1998) Parchment-skin illusion: sound-biased touch. Curr Biol 8:R190.

Kamphuisen A, Bauer M, van Ee R (2008) No evidence for widespread synchronized networks in binocular rivalry: MEG frequency tagging entrains primarily early visual cortex. J Vis 8:4 1-8.

Kayser C, Petkov CI, Augath M, Logothetis NK (2007) Functional imaging reveals visual modulation of specific fields in auditory cortex. J Neurosci 27:1824-1835.

Kayser C, Petkov CI, Remidios R, Logothetis NK (2012) Multisensory influences on auditory processing. In: The neural bases of multisensory processes (Murray MM, Wallace MT, eds). Boca Raton (FL): CRC Press.

Logothetis NK, Wandell BA (2004) Interpreting the BOLD signal. Annu Rev Physiol 66:735-769.

Melloni L, Molina C, Pena M, Torres D, Singer W, Rodriguez E (2007) Synchronization of neural activity across cortical areas correlates with conscious perception. J Neurosci 27:2858-2865.

Meredith MA, Stein BE (1986) Visual, auditory, and somatosensory convergence on cells in superior colliculus results in multisensory integration. J Neurophysiol 56:640-662.

Miller MW, Vogt BA (1984) Direct connections of rat visual cortex with sensory, motor, and association cortices. J Comp Neurol 226:184-202.

Molholm S, Ritter W, Murray MM, Javitt DC, Schroeder CE, Foxe JJ (2002) Multisensory auditory-visual interactions during early sensory processing in humans: a high-density electrical mapping study. Brain Res Cogn Brain Res 14:115-128.

Murray MM, Molholm S, Michel CM, Heslenfeld DJ, Ritter W, Javitt DC, Schroeder CE, Foxe JJ (2005) Grabbing your ear: rapid auditory-somatosensory multisensory interactions in low-level sensory cortices are not constrained by stimulus alignment. Cereb Cortex 15:963-974.

Neff DL, Green DM (1987) Masking produced by spectral uncertainty with multicomponent maskers. Percept Psychophys 41:409-415.

Noesselt T, Bonath B, Boehler CN, Schoenfeld MA, Heinze HJ (2008) On perceived synchrony-neural dynamics of audiovisual illusions and suppressions. Brain Res 1220:132-141.

Noesselt T, Rieger JW, Schoenfeld MA, Kanowski M, Hinrichs H, Heinze HJ, Driver J (2007) Audiovisual temporal correspondence modulates human multisensory superior temporal sulcus plus primary sensory cortices. J Neurosci 27:11431-11441.

Panagiotaropoulos TI, Deco G, Kapoor V, Logothetis NK (2012) Neuronal discharges and gamma oscillations explicitly reflect visual consciousness in the lateral prefrontal cortex. Neuron 74:924-935.

Pascual-Leone A, Walsh V (2001) Fast backprojections from the motion to the primary visual area necessary for visual awareness. Science 292:510-512.

Pins D, Ffytche D (2003) The neural correlates of conscious vision. Cereb Cortex 13:461-474.

Poppel E, Held R, Frost D (1973) Leter: Residual visual function after brain wounds involving the central visual pathways in man. Nature 243:295-296.

Raij T, Ahveninen J, Lin FH, Witzel T, Jaaskelainen IP, Letham B, Israeli E, Sahyoun C, Vasios C, Stufflebeam S, Hamalainen M, Belliveau JW (2010) Onset timing of cross-sensory activations and multisensory interactions in auditory and visual sensory cortices. Eur J Neurosci 31:1772-1782.

Regan D, Spekreijse H (1977) Auditory-visual interactions and the correspondence between perceived auditory space and perceived visual space. Perception 6:133-138.

Regan MP, Regan D (1988) A frequency domain technique for characterizing nonlinearities in biological systems. J theor Biol 133:293-317.

Regan MP, He P, Regan D (1995) An audio-visual convergence area in the human brain. Exp Brain Res 106:485-487.

Ress D, Heeger DJ (2003) Neuronal correlates of perception in early visual cortex. Nat Neurosci 6:414-420.

Rodriguez E, George N, Lachaux JP, Martinerie J, Renault B, Varela FJ (1999) Perception's shadow: long-distance synchronization of human brain activity. Nature 397:430-433.

Schlack A, Sterbing-D'Angelo SJ, Hartung K, Hoffmann KP, Bremmer F (2005) Multisensory space representations in the macaque ventral intraparietal area. J Neurosci 25:4616-4625.

Schroeder CE, Foxe JJ (2002) The timing and laminar profile of converging inputs to multisensory areas of the macaque neocortex. Brain Res Cogn Brain Res 14:187-198.

Schroeder CE, Lindsley RW, Specht C, Marcovici A, Smiley JF, Javitt DC (2001) Somatosensory input to auditory association cortex in the macaque monkey. J Neurophysiol 85:1322-1327.

Schroger E, Widmann A (1998) Speeded responses to audiovisual signal changes result from bimodal integration. Psychophysiology 35:755-759.

Sergent C, Baillet S, Dehaene S (2005) Timing of the brain events underlying access to consciousness during the attentional blink. Nat Neurosci 8:1391-1400.

Shams L, Seitz AR (2008) Benefits of multisensory learning. Trends Cogn Sci 12:411-417.

Shams L, Kamitani Y, Shimojo S (2000) Illusions. What you see is what you hear. Nature 408:788.

Shen D, Alain C (2011) Temporal attention facilitates short-term consolidation during a rapid serial auditory presentation task. Exp Brain Res 215:285-292.

Simons DJ, Levin DT (1997) Change blindness. Trends Cogn Sci 1:261-267.

Simons DJ, Rensink RA (2005) Change blindness: past, present, and future. Trends Cogn Sci 9:16-20.

Spence C, Squire S (2003) Multisensory integration: maintaining the perception of synchrony. Curr Biol 13:R519-521.

Stein BE, Stanford TR (2008) Multisensory integration: current issues from the perspective of the single neuron. Nat Rev Neurosci 9:255-266.

Strahan EJ, Spencer SJ, Zanna MP (2002) Subliminal priming and persuasion: Striking while the iron is hot. Journal of Experimental Social Psychology 38:556-568.

Sutoyo D, Srinivasan R (2009) Nonlinear SSVEP responses are sensitive to the perceptual binding of visual hemifields during conventional 'eye' rivalry and interocular 'percept' rivalry. Brain Res 1251:245-255.

Talsma D, Woldorff MG (2005) Selective attention and multisensory integration: multiple phases of effects on the evoked brain activity. J Cogn Neurosci 17:1098-1114.

Talsma D, Doty TJ, Woldorff MG (2007) Selective attention and audiovisual integration: is attending to both modalities a prerequisite for early integration? Cereb Cortex 17:679-690.

Talsma D, Senkowski D, Soto-Faraco S, Woldorff MG (2010) The multifaceted interplay between attention and multisensory integration. Trends Cogn Sci 14:400-410.

Teder-Salejarvi WA, McDonald JJ, Di Russo F, Hillyard SA (2002) An analysis of audio-visual crossmodal integration by means of event-related potential (ERP) recordings. Brain Res Cogn Brain Res 14:106-114.

Tong F, Meng M, Blake R (2006) Neural bases of binocular rivalry. Trends Cogn Sci 10:502-511.

Tsuchiya N, Koch C (2005) Continuous flash suppression reduces negative afterimages. Nat Neurosci 8:1096-1101.

van Atteveldt N, Formisano E, Goebel R, Blomert L (2004) Integration of letters and speech sounds in the human brain. Neuron 43:271-282.

Van der Burg E, Olivers CN, Bronkhorst AW, Theeuwes J (2008) Pip and pop: nonspatial auditory signals improve spatial visual search. J Exp Psychol Hum Percept Perform 34:1053-1065.

van Ee R, van Boxtel JJ, Parker AL, Alais D (2009) Multisensory congruency as a mechanism for attentional control over perceptual selection. J Neurosci 29:11641-11649.

Van Gijn J (2005) René Descartes (1596–1650). Journal of Neurology 252:241-242.

Vroomen J, Bertelson P, de Gelder B (2001) The ventriloquist effect does not depend on the direction of automatic visual attention. Percept Psychophys 63:651-659.

Weiskrantz L, Warrington EK, Sanders MD, Marshall J (1974) Visual capacity in the hemianopic field following a restricted occipital ablation. Brain 97:709-728.

117

Welch RB, Warren DH (1980) Immediate perceptual response to intersensory discrepancy. Psychol Bull 88:638-667.

Werner S, Noppeney U (2010) Distinct functional contributions of primary sensory and association areas to audiovisual integration in object categorization. J Neurosci 30:2662-2675.

Wiegand K, Gutschalk A (2012) Correlates of perceptual awareness in human primary auditory cortex revealed by an informational masking experiment. Neuroimage 61:62-69.

PERSONAL CONTRIBUTIONS

1. **Giani AS**, Conrad V, Watanabe M, Noppeney U (submitted) The invisible ventriloquist.

   Giani, Conrad and Noppeney designed the experiment. Stimuli were programmed and designed by Giani with the support of Conrad and Watanabe. Data was collected by Giani and Conrad with major support of Nathalie Christner and Beatrix Barth. Giani analyzed the data with help of Conrad and Noppeney. Giani and Noppeney wrote the manuscript.

2. **Giani AS**, Ortiz E, Belardinelli P, Kleiner M, Preissl H, Noppeney U (2011) Steady-state responses in MEG demonstrate information integration within but not across the auditory and visual senses. NeuroImage (60): 1478-1489.

   Giani and Noppeney designed the experiment. Stimuli were programmed and designed by Giani with the support of Kleiner. Giani collected and analyzed the data. Belardinelli, Ortiz, Preissl and Noppeney supported the analysis process. Giani and Noppeney wrote the manuscript.

3. **Giani AS**, Belardinelli P, Ortiz E, Kleiner M, Noppeney U (manuscript) From complex auditory scenes to perceptual awareness in 600 ms.

   Giani and Noppeney designed the experiment. Stimuli were programmed and designed by Giani with the support of Kleiner. Giani collected and analyzed the data. Belardinelli, Ortiz and Noppeney supported the analysis process. Giani and Noppeney wrote the manuscript.

ACKNOWLEDGEMENTS

To begin, I need to thank Uta Noppeney: With seemingly unlimited amounts of energy she supervised the current projects. In doing so, her various ideas, passion for new research avenues, mathematical skills and critical feedback profoundly shaped the outcome of this work.

During my PhD at the MPI I had the good fortune to be part of an extremely supportive research group. All my (former) colleagues (i.e. Tim Rohe, Verena Conrad, Panagiotis Tsiatsis, Ruth Adam, Hwee Ling Lee, Sebastian Werner, Sebastian van Saldern, Fabrizio Leo, Joana Leitão, Johannes Tuennerhoff, Michael Bannert and Remi Gau) supported or advised my work in one way or another. However, in particular, I would like to mention Joana, who spared no effort to assist me as a colleague and friend. Not officially part of the "Cognitive Neuroimaging" group, Mario Kleiner nevertheless regularly supported our work (and coffee breaks). His invaluable knowledge of computer hardware, in combination with the Psychtoolbox and many "fairy tales and horror stories", greatly added to the success of this work. I would also like to mention Masataka Watanabe. By means of his help we achieved a major breakthrough, when we were finally able to continuously suppress visual flashes. Similarly, I have to thank our "HiWis" Natalie Christner, Simone Götze and Beatrix Barth who tested many participants and helped fine-tuning parameters during for the ventriloquist experiment. Lastly, I have to thank Professor Heinrich Bülthoff who secured my funding even when my 'official' time as a PhD student was up.

Next to my colleagues at the MPI, I have to thank all my collaborators and friends at the MEG Center in Tübingen. In particular, I would like to mention Hubert Preissl and Christoph Braun who supported me whenever necessary and even beyond. Thanks a lot for your belief in my skills! Many, many thanks to Erick Ortiz and Paolo Belardinelli who advised and actively supported me with SPM and fieldtrip in general and source analyses in particular. I also have to thank Rossitza Draganova for valuable advices on steady-state responses and a nice time in Paris, Jürgen Dax for technical support, as well as Maike Borutta and Gabi Walker for MEG training and more practical support. I really enjoyed my time at the MEG Center and I thank all the people who (directly or indirectly) supported me and made me feel welcome.

This work would not have been possible without the continuous support and help of my dear parents Ruth and Herbert Giani. I am also extremely grateful to Tono Refeld who shared my joys and sorrows throughout the last years and who always cheers me up. Last but not least: Thanks to all 'Quilombolas' who made me feel at home in Tübingen.